W0106018

Nonlinear Ill–Posed Problems

APPLIED MATHEMATICS AND
MATHEMATICAL COMPUTATION

Editors

R.J. Knops and K.W. Morton

This series presents texts and monographs at graduate and research level covering a wide variety of topics of current research interest in modern and traditional applied mathematics, in numerical analysis and computation.

1 Introduction to the Thermodynamics of Solids *J.L. Ericksen* (1991)

2 Order Stars *A. Iserles and S.P. Nørsett* (1991)

3 Material Inhomogeneities in Elasticity *G. Maugin* (1993)

4 Bivectors and Waves in Mechanics and Optics *Ph. Boulanger and M. Hayes* (1993)

5 Mathematical Modelling of Inelastic Deformation *J.F. Besseling and E van der Geissen* (1993)

6 Vortex Structures in a Stratified Fluid: Order from Chaos *Sergey I. Voropayev and Yakov D. Afanasyev* (1994)

7 Numerical Hamiltonian Problems *J.M. Sanz-Serna and M.P. Calvo* (1994)

8 Variational Theories for Liquid Crystals *E. G. Virga* (1994)

9 Asymptotic Treatment of Differential Equations *A. Georgescu* (1995)

10 Plasma Physics Theory *A. Sitenko and V. Malnev* (1995)

11 Wavelets and Digital Signal Processing *A. Cohen and R.D. Ryan* (1995)

12 Numerical Solution of Convection-Diffusion Problems *K. W. Morton* (1996)

13 Weak and Measure-valued Solutions to Evolutionary PDEs
J. Málek, J. Nečas, M. Rokyta and M. Růžička

(Full details concerning this series, and more information on titles in preparation are available from the publisher.)

Nonlinear Ill–Posed Problems
(Volume 2)

A.N. TIKHONOV

The late founder and
Dean of the Faculty of Computational Mathematics
Cybernetics at the Moscow State University

A.S. LEONOV

Department of Mathematics
at the Moscow Physical Engineering Institute

and

A.G. YAGOLA

Department of Mathematics of the Physical Faculty
at Moscow State University

Springer Science+Business Media, B.V.

English language edition 1998

Originally published by Chapman & Hall in 1998
Softcover reprint of the hardcover 1st edition 1998

ISBN 978-94-017-5169-8 ISBN 978-94-017-5167-4 (eBook)
DOI 10.1007/978-94-017-5167-4

Nonlinear Ill–Posed Problems © Nauka
Original Russian language edition

Typeset in Latex by Focal Image Ltd., 20 Conduit Place, London W2 1HZ, UK

A catalogue record for this book is available from the British Library

Symbols

$\widehat{D}$ – finite-dimensional counterpart of the set D;

H – vector of the total error of an operator equation;

$H \equiv (h, 1/M, 1/N)$;

$\hat{I}(\hat{z}) \equiv \hat{I}_\eta(\hat{z}) \equiv f[\hat{J}_\eta(\hat{z})]$;

$\hat{J}_\eta(\hat{z}) \equiv \hat{J}_{\delta N}(\hat{z})$ – finite-dimensional approximate functional;

M – dimension of a finite-dimensional space U_M;

$\widehat{M}^\alpha[\hat{z}]$ – finite-dimensional smoothing functional;

N – dimension of a finite-dimensional space Z_N;

P_N and $\bar{P}_N$ – operators of coupling the spaces Z and Z_N;

$\bar{Q}_M$ – operator of coupling the spaces U_M and U;

U_M – finite-dimensional space approximating U;

$\hat{u}$ and $\hat{u}_{\sigma M}$ – finite-dimensional right-hand sides of an operator equation;

Z_N – finite-dimensional space approximating Z;

$\widehat{Z}^\alpha$ – set of extremals of a finite-dimensional smoothing functional for a given α;

$\hat{z}$ – element of the space Z_N;

$\hat{z}^\alpha$ – extremal of a finite-dimensional smoothing functional;

$\hat{z}_\eta$ – finite-dimensional approximate solution;

z_η – approximate solution;

$z_\eta = \bar{P}_N \hat{z}_\eta$;

$\hat{\beta}(\alpha)$, $\hat{\gamma}(\alpha)$, $\hat{\varepsilon}(\alpha)$, $\hat{\bar{\varepsilon}}(\alpha)$, $\hat{\pi}(\alpha)$, $\hat{\bar{\pi}}(\alpha)$, $\hat{\rho}(\alpha)$, $\hat{\bar{\rho}}(\alpha)$ – finite-dimensional counterparts of auxiliary functions;

η – vector of the approximation error of an extremal problem;

$\eta = (\delta, 1/N)$;

$\hat{\lambda} \equiv \hat{\lambda}_\eta$ – finite-dimensional generalized measure of incompatibility;

$\hat{\lambda}_{\varkappa}$ – $\varkappa$-approximation to $\hat{\lambda}$;

$\hat{\mu} \equiv \hat{\mu}_{\eta}$ – finite-dimensional approximate measure of incompatibility;

$\hat{\nu} \equiv \hat{\nu}_{\eta}$ – finite-dimensional counterpart of ν_{δ};

$\xi(1/M, \sigma)$ – measure of the finite-dimensional approximation of the right-hand side of an operator equation;

$\Psi(\eta, \Omega)$ – total measure of the approximation of a problem;

$\widehat{\Psi}(\eta, \Omega)$ – measure of the finite-dimensional approximation;

$\widehat{\psi}(H, \Omega)$ – measure of the finite-dimensional approximation of an approximate operator;

$\widehat{\Omega}(\hat{z})$ – finite-dimensional counterpart of a stabilizing functional;

$\widehat{\omega}_{\eta}$ – parameter of the finite-dimensional **g.p.q.** algorithm;

$\widehat{\Omega}^* \equiv \widehat{\Omega}^*_N$ – finite-dimensional counterpart of Ω^*;

A – $m \times n$-matrix;

$\bar{A}$ – exact matrix of a system of linear equations;

A_h – approximate matrix of a system;

$\widetilde{A}_h$ – matrix of the **m.p.m.** method;

$\widetilde{A}_h^+$ – minimal pseudoinverse matrix;

$\widehat{A}_h$ – special matrix of the **m.p.m.** method;

$\widehat{A}_{\lambda(h)} \equiv A_{\lambda(h)}$ – matrix of the **m.p.m.** principle;

E – unit $m \times m$-matrix;

h – error of an approximate matrix;

I – unit $n \times n$-matrix;

$\bar{u}$ – exact right-hand side of a system;

u_{σ} – approximate right-hand side of a system;

$x(\lambda)$ – a positive root to the equation $x^4 - x^3 = \lambda\,\rho^{-4}$;

$\bar{z}$ – normal pseudosolution of a system;

z_{δ} – approximation to $\bar{z}$;

z_{η} – approximate solution obtained by the **m.p.m.** method;

$\beta(\lambda)$ – discrepancy of the **m.p.m.** method;

η – vector of the error;

$\eta = (h, \sigma)$;

$\theta(x) = \{x^{-1} \text{if } x \neq 0;\ 0 \text{ if } x = 0\}$;

λ, $\lambda(h)$ – regularization parameter of the **m.p.m.** method;

$\nu(A)$ – condition number of the matrix A;

σ – error in specifying the right-hand side of a system;

$\mathfrak{A}_0$ – set of a priori restrictions on the matrix A;

$\mathfrak{A}$ – normed space of all $m \times n$-matrices;

$\mathfrak{A}^*$ – normed space of all $n \times m$-matrices;

$\mathfrak{A}_n$ – normed space of all $n \times n$-matrices;

$\mathfrak{A}_h = \{ A \in \mathfrak{A}_0 : \ \|A - A_h\| \leq h \}$.

CHAPTER 4

Finite-dimensional variants of algorithms

The main goal of our studies is to investigate finite-dimensional approximation of nonlinear ill-posed problems and, to assist the users, produce finite-dimensional variants of the algorithms which were considered in the first volume.

4.1 Finite-dimensional approximation of nonlinear ill-posed problems

Let (Z, τ) be a topological Hausdorff space. We assume that a functional $J(z)$ is defined on a subset $D \subseteq Z$, $D \neq \varnothing$. Consider the problem of minimizing the functional $J(z)$ on the set D in which it is required to obtain elements $z^* \in D$ such that

$$J(z^*) = \inf\{J(z)\colon z \in D\} \equiv J^* \tag{1}$$

Let problem (1) possess a nonempty set of solutions Z^*. Assuming this to be the case, the problem of finding Ω-optimal solutions consists of determining elements $\bar{z} \in Z^*$ for which

$$\Omega(\bar{z}) = \inf\{\Omega(z^*)\colon z^* \in Z^*\} \equiv \bar{\Omega}$$

In the general setting, the functional $\Omega(z)$ defined on the set D is supposed to be bounded below:

$$\inf\{\Omega(z)\colon z \in D\} \equiv \Omega^* > -\infty$$

and the functional $J(z)$ given in problem (1) statement in the explicit form is supposed to be unknown. For the approximate determination of Ω-optimal solutions we have at our disposal the data $\{J_\delta, \delta, \Psi_0\}$, where $J_\delta(z)$ is an approximate functional of problem (1) that belongs to the set of admissible approximate functionals $\mathcal{F}$, the elements of which are defined and bounded from below on D and satisfy the condition of approximation:

$$\left|J(z) - J_\delta(z)\right| \leq \Psi_0(\delta, \Omega(z)) \qquad \forall z \in D \tag{2}$$

Here the approximation measure Ψ_0 and the numerical vector $\delta \in \mathbf{R}^k_+$ $(\delta \neq \theta = (0, \ldots, 0) \in \mathbf{R}^k)$ are supposed to be known.

The fundamental problem of approximately determinining Ω-optimal solutions is to obtain elements $z_\delta \equiv z(J_\delta, \delta, \Psi_0) \in D$, which converge τ-sequentially to the set $\bar{Z}$ of Ω-optimal solutions of the problem (1) as $\|\delta\| \to 0$. That problem has been already considered in Chapter 2 among others. We assume that the functionals J, J_δ and Ω and the function Ψ_0 obey the requirements of Section 2.2 of Chapter 2.

As a rule, the numerical solution of the fundamental problem would be impossible without the use of computers. In general, the traditional way of implementation is to perform a finite-dimensional approximation of the problem under consideration. In many cases, the available data include finite-dimensional quantities which could be treated, in a certain sense, as approximate functionals J_δ. In this chapter, we study the problem of constructing finite-dimensional approximations to a solution of the fundamental problem under consideration by applying finite-dimensional variants of algorithms of the generalized principles of discrepancy, quasisolutions and smoothing functional (see Leonov (1988 b,c)).

We begin our exposition with a common setting of the problem of finite-dimensional approximation that makes it possible to draw fairly accurate outlines of the possible theory. Let Z_N be a normed space of the dimension N, the elements of which will be denoted by $\hat{z}_N$ (for example, $\hat{z}_N^* \in Z_N$). For the purposes of this chapter, we refer to operators $P_N\colon Z \to Z_N$ and $\bar{P}_N\colon Z_N \to Z$ possessing a number of remarkable properties:

(1) the operators P_N and $\bar{P}_N$ are continuous for all N;

(2) $P_N \bar{P}_N = E_N$, where E_N is the identity operator on the space Z_N;

(3) $\bar{P}_N P_N z \in D$ for any $z \in D$ and all positive integers N.

It is convenient to introduce

$$\widehat{D}_N \equiv \widehat{D} = \{\hat{z}_N\colon \bar{P}_N \hat{z}_N \in D\}$$

By virtue of property 3), the sets $\widehat{D}_N \subset Z_N$ appear to be nonempty because all elements of the form $\hat{z}_N = P_N z$ $(z \in D)$ are contained in them. As a matter of fact, the sets $\widehat{D}_N$ represent finite-dimensional counterparts of the set D. For brevity, the subscript N, which indicates the dimension of the corresponding finite-dimensional space Z_N, will now be omitted wherever it is held fixed in any convenient way (for example, we might agree to consider $\bar{P}\hat{z} \equiv \bar{P}_N \hat{z}_N$).

Suppose that instead of the functional $J_\delta(z)$ we deal with one of its finite-dimensional approximations, that is, a finite-dimensional functional of the form

$$\widehat{J}_{\delta N}(\hat{z}) \equiv \widehat{J}_\eta(\hat{z})$$

which is defined on the set $\widehat{D}_N$ and satisfies the condition of approximation:

$$\left|\widehat{J}_{\delta N}(P_N z) - J_\delta(z)\right| \le \widehat{\Psi}(\eta, \Omega(z)) \qquad \forall z \in D \tag{3}$$

Here the vector $\eta \in \mathbf{R}_+^{k+1}$ is of the form $\eta = (\delta, 1/N)$ and the function $\widehat{\Psi}$ is the measure of approximation of the functional J_δ by the finite-dimensional functional $\widehat{J}_\eta \equiv \widehat{J}_{\delta N}$. We assume the function $\widehat{\Psi}$ possesses the same properties as the function Ψ_0.

The fundamental finite-dimensional problem is to obtain from the available data $\{\widehat{J}_\eta, \widehat{\Psi}, \Psi_0, \delta, N, Z_N\}$ an element $\hat{z}_\eta \in \widehat{D}_N$ such that the family $\{z_\eta\}$, where $z_\eta \equiv \bar{P}_N \hat{z}_\eta \in D$, converges τ-sequentially to $\bar{Z}$ as $\eta \to 0$. Any such elements $z_\eta = \bar{P}_N \hat{z}_\eta$ will be adopted as solutions of the finite-dimensional problem at hand.

Let us describe in more detail the framework for solving the fundamental finite-dimensional problem by introducing on the set $\widehat{D}$ the functionals

$$\widehat{\Omega}(\hat{z}) \equiv \Omega(\bar{P}\hat{z}) \qquad \widehat{I}(\hat{z}) \equiv \widehat{I}_\eta(\hat{z}) = f[\widehat{J}_{\delta N}(\hat{z})]$$

where $f(x) \in \mathcal{F}^m$ is a fixed auxiliary function, and a finite-dimensional smoothing functional of the form

$$\widehat{M}^\alpha[\hat{z}] = \alpha\widehat{\Omega}(\hat{z}) + \widehat{I}(\hat{z}) \qquad \alpha > 0 \quad \hat{z} \in \widehat{D} \tag{4}$$

We may then set up the extremal problem for (4) in which it is necessary to obtain elements $\hat{z}^\alpha \in \widehat{D}$ such that

$$\widehat{M}^\alpha[\hat{z}^\alpha] = \inf\left\{\widehat{M}^\alpha[\hat{z}] : \hat{z} \in \widehat{D}\right\} \tag{5}$$

Let $\widehat{Z}^\alpha \neq \varnothing$ be the set of solutions for problem (5). To solve the fundamental finite-dimensional problem we are still in the framework of Section 2.4 of Chapter 2 consisting of two parts:

(a) making a substantiated choice of the regularization parameter $\alpha_\eta = \alpha\{\widehat{J}_\eta, \widehat{\Psi}, \Psi_0, \delta, N, Z_N\}$;

(b) selecting the solution $\hat{z}_\eta \equiv \hat{z}^{\alpha_\eta}$ of problem (5) for $\alpha = \alpha_\eta$ from the set $\widehat{Z}^{\alpha_\eta}$ in accordance with an approved rule, governing what can happen.

Various methods for choosing the regularization parameter and selecting the element $\hat{z}_\eta$ investigated in this chapter represent finite-dimensional variants of the algorithms developed in Sections 2.7–2.9 of Chapter 2. Along with the three conditions imposed above, we take for granted two additional conditions:

(4) $\overline{\lim}_{N\to\infty} \Omega(\bar{P}_N P_N \bar{z}) \leq \bar{\Omega}$ for any $\bar{z} \in \bar{Z}$;

(5) for each η the functional $\widehat{J}_{\delta N}$ is lower semicontinuous on the set $\widehat{D}$.

All five conditions provide a possibility to realize scheme (a)–(b) for solving the fundamental finite-dimensional problem. We need an auxiliary lemma to justify this approach.

Lemma 1 *For any fixed N, a nonempty set*

$$\widehat{\Omega}_C \equiv \{\hat{z} \in \widehat{D}_N : \widehat{\Omega}(\hat{z}) \leq C\}$$

is closed and compact in Z_N.

Proof. Consider an arbitrary sequence $\{\hat{z}_n\} \subset \widehat{\Omega}_C$ and the appropriate sequence of elements $\{z_n\}$, where $z_n = \bar{P}\hat{z}_n$. By the definition of the set $\widehat{D}$, the inclusion $\{z_n\} \subset D$ occurs with $\Omega(z_n) = \Omega(\bar{P}\hat{z}_n) = \widehat{\Omega}(\hat{z}_n) \leq C$. Thus, $\{z_n\} \subset \Omega_C = \{z \in D \colon \Omega(z) \leq C\}$. According to Assumption 2 of Section 2.2 of Chapter 2, the set Ω_C is τ-sequentially compact. Therefore, $\{z_n\}$ contains a subsequence $\{z_{n_k}\} \subset D$, which τ-converges as $k \to \infty$: $z_{n_k} \xrightarrow{\tau} z_0 \in \Omega_C$. From the continuity of the operator P and property 2) of the operators P and $\bar{P}$ we establish

$$Pz_{n_k} = P(\bar{P}\hat{z}_{n_k}) = \hat{z}_{n_k} \xrightarrow{Z_N} Pz_0 \equiv \hat{z}_0 \qquad \text{as } k \to \infty$$

Putting these together with continuity of the operator $\bar{P}$, we arrive at the chain of relations

$$z_{n_k} = \bar{P}\hat{z}_{n_k} \xrightarrow{\tau} \bar{P}\hat{z}_0$$

As far as (Z, τ) is a Hausdorff space, the sequence in question possesses a unique limit (see Section 1.1 in Chapter 1). We thus have $\bar{P}\hat{z}_0 = z_0 \in D$ and, hence, $\hat{z}_0 \in \widehat{D}$. The inclusion $z_0 \in \Omega_C$ implies the relations

$$\widehat{\Omega}(\hat{z}_0) = \Omega(\bar{P}\hat{z}_0) = \Omega(z_0) \leq C$$

which mean that $\hat{z}_0 \in \widehat{\Omega}_C$. □

Lemma 2 *For any fixed η the functionals $\widehat{\Omega}(\hat{z})$, $\widehat{I}(\hat{z})$, $\widehat{M^\alpha}[\hat{z}]$ are lower semicontinuous on $\widehat{D}$.*

Proof. We proceed as usual. This amounts to choosing an arbitrary sequence $\{\hat{z}_n\} \subset \widehat{D}$ such that $\hat{z}_n \to \hat{z}_0 \in \widehat{D}$ as $n \to \infty$. From the continuity of the operator $\bar{P}$ we deduce by definition of the set $\widehat{D}$ that

$$\bar{P}\hat{z}_n \xrightarrow{\tau} \bar{P}\hat{z}_0 \in D$$

as $n \to \infty$. Due to the τ-sequential lower semicontinuity of the functional Ω on the set D (see Section 2.2 in Chapter 2) we can derive the relations

$$\underline{\lim}_{\,n\to\infty}\widehat{\Omega}(\hat{z}_n) = \underline{\lim}_{\,n\to\infty}\Omega(\bar{P}\hat{z}_n) \geq \Omega(\bar{P}\hat{z}_0) = \widehat{\Omega}(\hat{z}_0)$$

From the lower semicontinuity of the functional $\widehat{J}_{\delta N}$ on the set $\widehat{D}$ and the continuity and increase of the function f it follows that

$$\underline{\lim}_{\,n\to\infty}\widehat{I}(\hat{z}_n) = \underline{\lim}_{\,n\to\infty} f[\widehat{J}_{\delta N}(\hat{z}_n)] \geq f[\widehat{J}_{\delta N}(\hat{z}_0)] = \widehat{I}(\hat{z}_0)$$

The preceding relations combined with equality (4) complete the proof of the lemma. □

Lemmas 1 and 2 confirm the validity of Assumptions 1–2 of Section 2.2 in Chapter 2 for the functionals $\widehat{\Omega}(\hat{z})$ and $\widehat{I}(\hat{z})$ on the set $\widehat{D}$ for any fixed numbers N and δ. In so doing, the topology of convergence in norm is accepted as a proper topology in Z_N.

Thus, Lemmas 1 and 2 imply a number of statements similar to those established in Sections 2.5–2.6 in Chapter 2 under Assumptions 1–2 of Section 2.2 in Chapter 2. In particular, the following assertion is valid.

Theorem 1 *For any fixed η and any $\alpha > 0$ the extremal problem* (5) *is solvable, so that $\widehat{Z}^\alpha \neq \varnothing$.*

Proof. The proof of the theorem can be carried out by applying Theorem 2.4.2 to the functional $\widehat{M}^\alpha$ and using the results of Lemmas 1 and 2. □

Let us stress, turning once again to problem (5), that we are dealing with a finite-dimensional problem. There are many effective methods for solving similar problems (Fiacco and McCormick (1968), Himmelblau (1971), Karmanov (1975), Polak (1971), Polyak (1974), Pshenichny (1993), Pshenichny and Danilin (1975), Vasilyev (1980)). The situation when $\widehat{D}$ is a convex set and $\widehat{\Omega}$, $\widehat{I}$ are convex functionals on $\widehat{D}$ is of particular convenience. In each such case, problem (5) falls within the category of problems relating to convex programming. When $\widehat{D} = Z_N$ and $\widehat{\Omega}$ and $\widehat{I}$ are quadratic functionals, problem (5) reduces to solving a system of linear equations (Karmanov (1975), Pshenichny and Danilin (1975), Vasilyev (1980)).

In conclusion we quote some corollaries to the approximation conditions (2)–(3). First, for any $z \in D$

$$\left|J(z) - \widehat{J}_\eta(Pz)\right| \leq \Psi_0(\delta, \Omega(z)) + \widehat{\Psi}(\eta, \Omega(z)) \equiv \Psi(\eta, \Omega(z)) \qquad (6)$$

By virtue of property 3) of the operators P and $\bar{P}$, we obtain $\bar{P}(P\bar{z}) \in D$ for any $\bar{z} \in \bar{Z} \subset D$. This means that the element $\hat{\bar{z}} \equiv P\bar{z}$ belongs to the set $\widehat{D}$. Therefore, it follows from (6) that

$$\left|J^* - \widehat{J}_\eta(\hat{\bar{z}})\right| = \left|J(\bar{z}) - \widehat{J}_\eta(P\bar{z})\right| \leq \Psi(\eta, \Omega(\bar{z})) = \Psi(\eta, \bar{\Omega}) \qquad (7)$$

Finally, in light of property 2) of the operators P and $\bar{P}$, we deduce from (6) that for each $\hat{z} \in \widehat{D}$

$$\left|J(\bar{P}\hat{z}) - \widehat{J}_\eta(P\bar{P}\hat{z})\right| = \left|J(\bar{P}\hat{z}) - \widehat{J}_\eta(\hat{z})\right| \leq \Psi(\eta, \widehat{\Omega}(\hat{z})) \qquad (8)$$

The function $\Psi(\eta, \Omega(z))$ involved in (6) represents the total measure of approximation of the exact functional J by the finite-dimensional functional $\widehat{J}_\eta$. Clearly, it possesses the same properties as the function Ψ_0 (see Section 2.2 in Chapter 2).

Remark Finite-dimensional approximations of ill-posed problems have been extensively investigated by many authors. In this regard, we would like to point out the papers and monographs by Ageev and Vasin (1979), Budak and Berkovich (1971), Budak *et al.* (1969), Gaponenko (1980, 1989), Ivanov *et al.* (1978), Liskovets (1981), Morozov (1974 a, 1984), Tanana (1975 a, b, 1981), Tikhonov and Arsenin (1977), Tikhonov and Glasko (1964, 1965), Tikhonov *et al.* (1983), Vasin (1972, 1975, 1979), Vasin and Tanana (1974). Sometimes the error of a finite-dimensional approximation arising when choosing a regularized approximation, especially under *a posteriori* choices, has been ignored. The accurate account of the error of a finite-dimensional approximation by the generalized principle of discrepancy was first made by Goncharskiĭ *et al.* (1974 a) in the study of linear operator equations. The papers of Eremin and Leonov (1975) and Leonov (1975, 1976, 1988 b, c) were devoted to this subject. In particular, the category of nonlinear ill-posed problems was considered by Leonov (1975) with alternative choice of regularization parameter. Leonov (1988 b, c) explored finite-dimensional approximation of nonlinear problems on the basis of the **g.p.d.**, **g.p.q.** and **g.p.f.** algorithms.

4.2 Finite-dimensional generalized principle of discrepancy

By fixing $\eta = (\delta, 1/N)$, we now consider

$$\begin{aligned}\hat{\lambda} &\equiv \hat{\lambda}_\eta = \inf\{\widehat{J}_\eta(\hat{z}) + \Psi(\eta, \widehat{\Omega}(\hat{z})) : \hat{z} \in \widehat{D}\} \\ \hat{\mu} &\equiv \hat{\mu}_\eta = \inf\{\widehat{J}_\eta(\hat{z}) : \hat{z} \in \widehat{D}\} \\ \widehat{\Omega}^* &\equiv \widehat{\Omega}^*_N = \inf\{\widehat{\Omega}(\hat{z}) : \hat{z} \in \widehat{D}\} = \inf\{\Omega(\bar{P}\hat{z}) : \hat{z} \in \widehat{D}\}\end{aligned}$$

It is straightforward to verify that $\hat{\lambda} \geq \hat{\mu}$ and $\widehat{\Omega}^* \geq \Omega^*$. Moreover, the numbers $\hat{\lambda}$ possess one useful property.

Lemma 1 *For any fixed η the relations $\hat{\lambda}_\eta \geq J^*$ and $\hat{\lambda}_\eta \to J^*$ take place as $\eta \to 0$.*

Proof. In light of property 2) of the operators P and $\bar{P}$, it follows from inequality (1.6) in Section 4.1 that

$$\begin{aligned}J^* &= \inf\{J(z) : z \in D\} \\ &\leq J(\bar{P}\hat{z}) \leq \widehat{J}_\eta(P\bar{P}\hat{z}) + \Psi(\eta, \Omega(\bar{P}\hat{z})) \\ &= \widehat{J}_\eta(\hat{z}) + \Psi(\eta, \widehat{\Omega}(\hat{z})) \quad \forall \hat{z} \in \widehat{D}\end{aligned} \tag{1}$$

Here, we have also taken into account that $\bar{P}\hat{z} \in D$ according to the definition of the set $\widehat{D}$. We infer from (1) and (1.7) that

$$\begin{aligned}J^* &\leq \inf\{\widehat{J}_\eta(\hat{z}) + \Psi(\eta, \widehat{\Omega}(\hat{z})) : \hat{z} \in \widehat{D}\} \\ &= \hat{\lambda}_\eta \leq \widehat{J}_\eta(\hat{\bar{z}}) + \Psi(\eta, \widehat{\Omega}(\hat{\bar{z}})) \\ &\leq J^* + \Psi(\eta, \bar{\Omega}) + \Psi(\eta, \Omega(\bar{P}_N P_N \bar{z}))\end{aligned} \tag{2}$$

By property 4) of the operators P and $\bar{P}$, we derive for all $N \geq N_0 =$ const the estimate $\Omega(\bar{P}_N P_N \bar{z}) \leq \widetilde{\Omega} =$ const. In turn, the properties of the function Ψ imply the relation

$$0 \leq \Psi(\eta, \Omega(\bar{P}_N P_N \bar{z})) \leq \Psi(\eta, \widetilde{\Omega}) \to 0 \qquad \eta \to 0$$

Thus, the convergences $\Psi(\eta, \Omega(\bar{P}_N P_N \bar{z})) \to 0$ and $\Psi(\eta, \bar{\Omega}) \to 0$ occur as

$\eta = (\delta, 1/N) \to 0$. Hence, passing in (2) to the limit as $\eta \to 0$, we establish the desired convergence $\hat{\lambda}_\eta \to J^*$. □

It is worth noting here that Lemma 1 is analogous to Theorem 2.3.1, but does not follow directly from it. Let us introduce the set

$$\widehat{Z}_0 \equiv \{\hat{z} \in \widehat{D} \colon \widehat{\Omega}(\hat{z}) = \widehat{\Omega}^*\}$$

Applying the results of Lemma 2.5.12 to the functionals $\widehat{\Omega}$ and $\widehat{I}$ with regard to Lemmas 1.1–1.2 yields that the set $\widehat{Z}_0$ is nonempty and there exist elements $\hat{z}_0 \in \widehat{Z}_0$ for which

$$\widehat{I}(\hat{z}_0) = \inf\left\{\widehat{I}(\hat{z}) \colon \hat{z} \in \widehat{Z}_0\right\} \equiv \hat{\nu}_\eta \equiv \hat{\nu}$$

Before going further, it is necessary to say that the extremal problem so formulated is analogous to problem (2.5.34).

In complete agreement with the standard framework (see Section 2.6 of Chapter 2), a first step is to introduce the auxiliary functions

$$\hat{\gamma}(\alpha) = \widehat{\Omega}(\hat{z}^\alpha) \qquad \hat{\beta}(\alpha) = \widehat{I}(\hat{z}^\alpha)$$
$$\hat{\varphi}(\alpha) = \widehat{M^\alpha}[\hat{z}^\alpha] \qquad \forall \hat{z}^\alpha \in \widehat{Z}^\alpha$$

From Section 2.6 of Chapter 2 and Lemmas 1.1–1.2, we know that those functions possess the properties established in Lemmas 2.6.1–2.6.6 and their corollaries. In what follows, the quantities J_δ^*, Ω^* and ν_δ of Section 2.6 in Chapter 2 have to be replaced by their finite-dimensional counterparts $\hat{\mu}$, $\widehat{\Omega}^*$ and $\hat{\nu}$. In addition, assume that, instead of the number $\hat{\lambda}$, its upper estimate is known with some degree of accuracy. More specifically, we have at our disposal a number $\hat{\lambda}_\varkappa \equiv \hat{\lambda}_{\eta\varkappa}$ such that $0 \le \hat{\lambda}_\varkappa - \hat{\lambda} \le \varkappa$. As before, it is reasonable to appeal to counterparts of functions (2.7.1)–(2.7.2):

$$\hat{\pi}(\alpha) = f[\hat{\lambda}_\varkappa + \Psi(\eta, \widehat{\Omega}(\hat{z}^\alpha))] \equiv \widehat{\Pi}[\hat{z}^\alpha] \qquad \hat{\rho}(\alpha) = \hat{\beta}(\alpha) - \hat{\pi}(\alpha) \equiv \widehat{P}[\hat{z}^\alpha]$$

Obviously, those functions possess the properties listed in Theorem 2.7.1 if we replace the numbers J_δ^*, λ_δ, ν_δ and Ω^* by their finite-dimensional counterparts.

Our starting point is the algorithm of the **finite-dimensional generalized principle of discrepancy (f.d.g.p.d.)**.

By a proper choice of the regularization parameter α_ζ ($\zeta \equiv (\eta, \varkappa)$) in the **f.d.g.p.d.** algorithm, we mean a solution of the equation with a monotone function

$$\hat{\rho}(\alpha) = 0 \tag{3}$$

From Lemma 2.6.7 and Theorems 2.7.2 and 2.7.4 we deduce the following assertion.

Theorem 1 *Let $\hat{\rho}_\infty \equiv \hat{\rho}(+\infty) > 0$ for a fixed ζ. Then equation* (3) *has at least one solution $\alpha_\zeta \ge 0$. If, in addition, $\hat{\rho}_0 \equiv \hat{\rho}(+0) < 0$, then any solution* *to equation* (3) *is positive. Let under the conditions* $\hat{\rho}_0 < 0$ *and* $\hat{\rho}_\infty > 0$ *the problem of minimizing the smoothing functional* (1.5) *has, for each* $\alpha > 0$, *a unique solution* $\hat{z}^\alpha$ *such that* $\hat{z}^{\alpha_1} \neq \hat{z}^{\alpha_2}$ *for any* α_1 *and* α_2 $(0 < \alpha_1 < \alpha_2)$. *In this case, equation* (3) *possesses a unique ordinary solution* $\alpha_\zeta > 0$.

To understand the results obtained a little better, we describe in the next theorem sufficient conditions for the validity of the inequalities $\hat{\rho}_0 \equiv \hat{\rho}_0^\zeta < 0$ and $\hat{\rho}_\infty \equiv \hat{\rho}_\infty^\zeta > 0$.

Theorem 2 *Let the conditions in the general setting of Section* 1 *be accompanied by the following ones:*

(a) $Z^* \cap Z_0 = \varnothing$;

(b) at least one of the functions $\Psi_0(\delta, \Omega)$, $\widehat{\Psi}(\eta, \Omega)$ increases in the second argument when δ ($\|\delta\| \neq 0$) and N are held fixed.

Then for all sufficiently small $\|\eta\|$ and $\varkappa$ $(0 < \|\eta\| \leq \Delta_0, 0 \leq \varkappa \leq \varkappa_0)$ the inequalities $\hat{\rho}_0^\zeta < 0$ and $\hat{\rho}_\infty^\zeta > 0$ are valid.

Proof. For the most part, the proof follows the scheme of Theorem 2.7.3. Therefore, we are only interested in some features of the proof. First they are connected with the establishment of the convergence $\hat{\nu}_\eta \to \nu_0$ as $\eta \to 0$.

We look for $z_0 \in Z_0$ in just the same way as in Theorem 2.7.3:

$$\nu_0 = \inf\left\{f[J(z)] : z \in Z_0\right\} = f[J(z_0)]$$

and choose an element $\hat{z}_0 \in \widehat{Z}_0$ subject to the condition

$$\hat{\nu} = \inf\left\{f[\widehat{J}_\eta(\hat{z})] : \hat{z} \in \widehat{Z}_0\right\} = f[\widehat{J}_\eta(\hat{z}_0)]$$

Using inequalities (1.8) and (1.6), we obtain

$$\begin{aligned} \nu_0 = f[J(z_0)] \leq f[J(\bar{P}\hat{z}_0)] &\leq f[\widehat{J}_\eta(\hat{z}_0) + \Psi(\eta, \widehat{\Omega}(\hat{z}_0))] \\ &= f[\widehat{J}_\eta(\hat{z}_0) + \Psi(\eta, \widehat{\Omega}^*)] \\ \hat{\nu} = f[\widehat{J}_\eta(\hat{z}_0)] \leq f[\widehat{J}_\eta(Pz_0)] &\leq f[J(z_0) + \Psi(\eta, \Omega(z_0))] \\ &= f[J(z_0) + \Psi(\eta, \Omega^*)] \end{aligned}$$

By analogy with (2.7.5) the latter inequalities imply

$$f[f^{-1}(\nu_0) - \Psi(\eta, \widehat{\Omega}^*)] \leq \hat{\nu}_\eta \leq f[f^{-1}(\nu_0) + \Psi(\eta, \Omega^*)] \tag{4}$$

By the definition of $\widehat{\Omega}^*$,

$$\widehat{\Omega}_N^* \leq \widehat{\Omega}(P_N \bar{z}) = \Omega(\bar{P}_N P_N \bar{z})$$

Thus, in light of property 4) of the operators P and $\bar{P}$, the quantities $\widehat{\Omega}_N^*$ are bounded. Due to the monotonicity of the function $\Psi(\eta, \Omega)$ in the second argument and the convergence $\Psi(\eta, \Omega) \to 0$, as $\eta \to 0$, we establish

$$\Psi(\eta, \widehat{\Omega}_N^*) \to 0 \quad \text{and} \quad \Psi(\eta, \Omega^*) \to 0$$

In view of relations (4), the function f being continuous provides $\hat{\nu}_\eta \to \nu_0$ as $\eta \to 0$.

Combination of the preceding convergence, Lemma 1 and Theorem 2.7.1 gives

$$\begin{aligned}\lim_{\zeta\to 0} \hat{\rho}^\zeta_\infty &= \lim_{\eta\to 0} \hat{\nu}_\eta - \lim_{\eta,\varkappa\to 0} f[\Psi(\eta, \widehat{\Omega}^*_N) + \hat{\lambda}_\varkappa] \\ &= \nu_0 - \lim_{\eta,\varkappa\to 0} f[\hat{\lambda}_\varkappa] \\ &= \nu_0 - f(J^*) \equiv \nu_0 - \mu_0 > 0\end{aligned}$$

Here we used also the inequality of Theorem 2.7.3 $\nu_0 > \mu_0$ arising from the condition $Z^* \cap Z_0 = \varnothing$. On the basis of the inequality

$$\lim_{\zeta\to 0} \hat{\rho}^\zeta_\infty > 0$$

it is possible to deduce, as we did in Theorem 2.7.3, that $\hat{\rho}^\zeta_\infty > 0$ for $0 < \|\eta\| \le \Delta_0$ and $\varkappa \le \varkappa_0$.

The proof of the inequality $\hat{\rho}^\zeta_0 > 0$ is carried out in a similar way as in Theorem 2.7.3 by replacing the quantities Ω_δ, Ω^*, λ_δ and δ by their counterparts $\widehat{\Omega}_\eta \equiv \hat{\gamma}(+0)$, $\widehat{\Omega}^*$, $\hat{\lambda}_\varkappa$ and ζ. Here we must take into account the monotone increase of the function $\Psi(\eta, \Omega)$ in the second argument. This property is an immediate implication of condition (b) of the theorem. □

Let the constants $q > 1$ and $C > 1$ be known in advance and let $\alpha_\zeta > 0$ be obtained by means of the **f.d.g.p.d.** algorithm. In dealing with $\alpha_1 \equiv \alpha_\zeta/q$ and $\alpha_2 \equiv \alpha_\zeta q$ we consider two arbitrary solutions $\hat{z}^{\alpha_1}$ and $\hat{z}^{\alpha_2}$ of problem (1.5) corresponding to the values $\alpha = \alpha_1$ and $\alpha = \alpha_2$, respectively.

We begin by placing the selection rule of the **f.d.g.p.d.** algorithm.

Selection rule *If the inequality*

$$\widehat{I}(\hat{z}^{\alpha_2}) \ge C\widehat{\Pi}(\hat{z}^{\alpha_1}) - (C-1)f(\hat{\lambda}_\varkappa) \tag{5}$$

holds for a fixed $\zeta = (\delta, 1/N, \varkappa)$, then one can adopt as a solution of the fundamental finite-dimensional problem complying with available data $(\widehat{J}_\eta, \Psi_0, \widehat{\Psi}, \delta, N, \varkappa)$ an element $z_\zeta = \bar{P}_N \hat{z}_\zeta$, where $\hat{z}_\zeta \equiv \hat{z}^{\alpha_\zeta}$ is chosen from the set $\widehat{Z}^{\alpha_\zeta}$ in such a way to satisfy the inequality

$$\widehat{P}[\hat{z}^{\alpha_\zeta}] = \widehat{I}(\hat{z}^{\alpha_\zeta}) - \widehat{\Pi}(\hat{z}^{\alpha_\zeta}) \le 0 \tag{6}$$

For example, one can take $\hat{z}_\zeta = \hat{z}^{\alpha_\zeta}_-$ (see Lemma 2.6.5).
In the case where

$$\widehat{I}(\hat{z}^{\alpha_2}) \le C\widehat{\Pi}(\hat{z}^{\alpha_1}) - (C-1)f(\hat{\lambda}_\varkappa) \tag{7}$$

one can accept as a solution of the fundamental finite-dimensional problem an element $z_\zeta = \bar{P}_N \hat{z}^{\alpha_\zeta}$, where $\hat{z}^{\alpha_\zeta}$ is chosen from the set $\widehat{Z}^{\alpha_\zeta}$ subject to the condition

$$\widehat{P}[\hat{z}^{\alpha_\zeta}] \ge 0 \tag{8}$$

For example, the choice $\hat{z}^{\alpha_\zeta} = \hat{z}_+^{\alpha_\zeta}$ suits us perfectly.

In this context, convergence of approximate solutions will be of great interest.

Theorem 3 *If for an arbitrary sequence $\{\zeta_n\}$, $\zeta_n = (\delta_n, 1/N_n, \varkappa_n)$ converging to 0 as $n \to \infty$ the sequence $\{z_n\}$, $z_n \equiv z_{\zeta_n} = \bar{P}_{N_n}\hat{z}^{\alpha_{\zeta_n}}$, is obtained by means of the* **f.d.g.p.d.** *algorithm, then*

$$z_n \xrightarrow{T} \bar{Z} \text{ and } \Omega(z_n) \to \hat{\Omega} \text{ as } n \to \infty$$

Proof. For conditions (a)–(e) of Section 2.5 in Chapter 2 to be valid with respect to the families $\{\bar{P}\hat{z}^{\alpha_\zeta}\}$ and $\{\bar{P}\hat{z}^{\alpha_{1,2}}\}$, the role of the numbers $\alpha(\delta)$ and λ_δ will be played by the quantities α_ζ and $\hat{\lambda}_\varkappa$. To expound certain exploratory devices for establishing them, we introduce the functionals $J_\eta(\bar{P}\hat{z}) \equiv \hat{J}_\eta(P\bar{P}\hat{z}) = \hat{J}_\eta(\hat{z})$ and $I_\eta(\bar{P}\hat{z}) \equiv f[J_\eta(\bar{P}\hat{z})] = f[\hat{J}_\eta(\hat{z})] = \hat{I}(\hat{z})$ defined for any $\hat{z} \in \hat{D}$. It is easily seen from (1.7) that

$$\left|J^* - J_\eta(\bar{P}\hat{\bar{z}})\right| \le \Psi(\eta, \bar{\Omega}) \tag{9}$$

For verifying condition (a) of Section 2.5 in Chapter 2 we use the extremality of the element $\hat{z}^{\alpha_\zeta}$, being a solution of problem (1.5), inequality (9) and the increase of the function f:

$$\begin{aligned}
\widehat{M}^{\alpha_\zeta}[\hat{z}^{\alpha_\zeta}] &= \alpha_\zeta\hat{\Omega}(\hat{z}^{\alpha_\zeta}) + \hat{I}(\hat{z}^{\alpha_\zeta}) = \alpha_\zeta\Omega(\bar{P}\hat{z}^{\alpha_\zeta}) + I_\eta(\bar{P}\hat{z}^{\alpha_\zeta}) \\
&\le \widehat{M}^{\alpha_\zeta}[\bar{P}\bar{z}] = \alpha_\zeta\Omega(\bar{P}P\bar{z}) + f[J_\eta(\bar{P}\hat{\bar{z}})] \\
&\le \alpha_\zeta\Omega(\bar{P}P\bar{z}) + f[J^* + \Psi(\eta, \bar{\Omega})] \\
&\le \alpha_\zeta\bar{\Omega}_\eta + f[J^* + \Psi(\eta, \bar{\Omega}_\eta)]
\end{aligned} \tag{10}$$

accepting $\bar{\Omega}_\eta \equiv \max\{\bar{\Omega}, \Omega(\bar{P}_N P_N \bar{z})\}$ and taking into account that $\bar{\Omega}_\eta \ge \bar{\Omega}$. As a matter of fact, inequality (10) represents condition (a) with $\bar{\Omega}_\delta = \bar{\Omega}_\eta$, since condition (4) related to the operators $\bar{P}_N$ and P_N provides the convergence $\bar{\Omega}_\eta \to \bar{\Omega}$ as $\eta = (\delta, 1/N) \to 0$.

Likewise, we can deduce condition (b) by taking the form

$$\widehat{M}^{\alpha_2}[\hat{z}^{\alpha_2}] = \alpha_2\Omega(\bar{P}\hat{z}^{\alpha_2}) + I_\eta(\bar{P}\hat{z}^{\alpha_2}) \le \alpha_\zeta q\bar{\Omega}_\eta + f[J^* + \Psi(\eta, \bar{\Omega}_\eta)]$$

Condition (c), connected to the monotonicity of the auxiliary functions $\hat{\gamma}$ and $\hat{\beta}$, follows from Lemma 2.6.1 and the inequality $\alpha_1 < \alpha_\zeta < \alpha_2$:

$$\begin{aligned}
\Omega(\bar{P}\hat{z}^{\alpha_1}) &= \hat{\Omega}(z^{\alpha_1}) = \hat{\gamma}(\alpha_1) \ge \hat{\gamma}(\alpha_\zeta) = \Omega(\bar{P}\hat{z}^{\alpha_\zeta}) \ge \hat{\gamma}(\alpha_2) = \Omega(\bar{P}\hat{z}^{\alpha_2}) \\
I_\eta(\bar{P}\hat{z}^{\alpha_1}) &= \hat{I}(\hat{z}^{\alpha_1}) = \hat{\beta}(\alpha_1) \le \hat{\beta}(\alpha_\zeta) = I_\eta(\bar{P}\hat{z}^{\alpha_\zeta}) \le \hat{\beta}(\alpha_2) = I_\eta(\bar{P}\hat{z}^{\alpha_2})
\end{aligned}$$

Condition (e) can be justified in a similar way. Finally, condition (d) holds true thanks to the selection rule (5)–(8) approved for the regularization parameter.

The fulfilment of conditions (a)–(e) implies the validity of Lemmas 2.5.7–2.5.10 for $\Omega(\bar{P}\hat{z}^{\alpha_\zeta})$ and $I_\eta(\bar{P}\hat{z}^{\alpha_\zeta})$. In turn, we stated in Section 2.5 in

Chapter 2 that Lemmas 2.5.7–2.5.10 yield

$$\overline{\lim}_{\zeta\to 0}\Omega(\bar{P}_N\hat{z}^{\alpha_\zeta}) \le \bar{\Omega} \qquad \lim_{\zeta\to 0} I_\eta(P_N\hat{z}^{\alpha_\zeta}) = I^* = f(J^*)$$

Due to Lemma 2.5.1 and Corollary 2.5.1, the preceding relations guarantee the required convergences: $z_n \xrightarrow{T} \bar{Z}$ and $\Omega(z_n) \to \bar{\Omega}$ as $n \to \infty$. □

Theorem 3 justifies that the **f.d.g.p.d.** algorithm gives a solution of the fundamental finite-dimensional problem and is a regularizing algorithm for solving the variational problem (1.1).

4.3 Finite-dimensional generalized principles of quasisolutions and smoothing functional

The finite-dimensional generalized principle of quasisolutions (f.d.g.p.q.) is implemented by means of available data $\{\widehat{J}_\eta, \Psi_0, \widehat{\Psi}, \delta, N, Z_N\}$ and the quantity $\hat{\omega}_\eta$ being an estimator of $\bar{\Omega}$.

The problem of making a substantiated choice of the regularization parameter α_η in the **f.d.g.p.q.** algorithm is to solve the equation with a monotone function

$$\hat{\gamma}(\alpha) = \hat{\omega}_\eta \tag{1}$$

From Lemma 2.6.7 and Theorems 2.8.1, 2.8.2 we obtain the following assertion.

Theorem 1 *If for any fixed η the inequality $\widehat{\Omega}^* < \hat{\omega}_\eta < \widehat{\Omega}_\eta$ holds, then equation* (1) *has at least one solution $\alpha_\eta > 0$. More specifically, if the extremal problem* (1.5) *possesses, for each $\alpha > 0$, a unique solution $\hat{z}^\alpha$ and $\hat{z}^{\alpha_1} \neq \hat{z}^{\alpha_2}$ for any α_1 and α_2 $(0 < \alpha_1 < \alpha_2)$, then equation* (1) *admits a unique ordinary solution.*

Assuming that $\alpha_\eta > 0$ is a solution to equation (1), let us formulate the selection rule of the **f.d.g.p.q.** algorithm. Given constants $C > 1$ and $q > 1$, set $\alpha_1 \equiv \alpha_\eta/q$ and $\alpha_2 \equiv \alpha_\eta q$. Let $\hat{z}^{\alpha_1}$ and $\hat{z}^{\alpha_2}$ be arbitrary solutions of problem (1.5) for $\alpha = \alpha_1$ and $\alpha = \alpha_2$, respectively.

Selection rule *When inequality (2.5) holds true, an element $\hat{z}^{\alpha_\eta}$ of the set $\widehat{Z}^{\alpha_\eta}$ is chosen so that $\widehat{\Omega}(\hat{z}^{\alpha_\eta}) \ge \hat{\omega}_\eta$. For example, one can take $\hat{z}^{\alpha_\eta} = \hat{z}_-^{\alpha_\eta}$.*

Provided condition (2.7) holds, we may choose from the set $\widehat{Z}^{\alpha_\eta}$ an element $\hat{z}^{\alpha_\eta}$ such that $\widehat{\Omega}(\hat{z}^{\alpha_\eta}) \le \hat{\omega}_\eta$. For example, $\hat{z}^{\alpha_\eta} = \hat{z}_+^{\alpha_\eta}$ is good enough for our purposes.

Observe that the element $\hat{z}^{\alpha_\eta}$ thus obtained depends on the number $\hat{\lambda}_\varkappa$ (see (2.5) and (2.7)), that is, on $\varkappa$. In other words, the preceding selection rule involves $\hat{z}^{\alpha_\eta} = \hat{z}^{\alpha_\eta}(\varkappa)$.

With this in mind, one can adopt $z_\zeta = \bar{P}_N\hat{z}^{\alpha_\eta}(\varkappa)$ as a solution of the fundamental finite-dimensional problem under consideration.

Theorem 2 *Let numbers $\hat{\omega}_\eta$ satisfy the relations $\hat{\omega}_\eta \geq \bar{\Omega}$ and $\hat{\omega}_\eta \to \bar{\Omega}$ as $\eta \to 0$. Assume that, for a sequence $\{\zeta_n\}$ converging to 0 as $n \to \infty$, the quantities $\alpha_{\eta_n} > 0$ and $\hat{z}^{\alpha_{\eta_n}}(\varkappa_n)$ are obtained by means of the* **f.d.g.p.q.** *algorithm. Then for any sequence $\{z_n\}$, $z_n = \bar{P}_{N_n}\hat{z}^{\alpha_{\eta_n}}(\varkappa_n)$, the limit relations $z_n \xrightarrow{T} \bar{Z}$ and $\Omega(z_n) \to \bar{\Omega}$ occur as $n \to \infty$.*

Proof. The proof is carried out by the same scheme as Theorem 2.3 on the basis of Lemma 2.5.11 and Corollary 2.5.1 with respect to the family $\{P_N \hat{z}^{\alpha_\eta}(\varkappa)\}$. Conditions (a)–(e) of Section 2.5 in Chapter 2 can be checked in just the same way as in Theorem 2.3 by replacing the numbers α_ζ by α_η. The modification of conditions (d) from Lemma 2.5.11 is taken into account in contents of the selection rule of the **f.d.g.p.q.** algorithm. □

Thus, by Theorem 2, the family $\{\bar{P}_N \hat{z}^{\alpha_\eta}(\varkappa)\}$ constitutes what is called a solution of the fundamental finite-dimensional problem.

Our next step is the **finite-dimensional generalized principle of smoothing functional (f.d.g.p.s.f.)**. For the same reason as above we are concerned with a new function

$$\hat{\varepsilon}(\alpha) \equiv \hat{\varphi}(\alpha) - f[\hat{\lambda}_\varkappa + \Psi^p(\eta, \widehat{\Omega}(\hat{z}^\alpha))] \equiv \widehat{E}(\hat{z}^\alpha) \qquad \alpha > 0$$

where a fixed parameter p is taken from $0 < p < 1$. The function $\hat{\varepsilon}(\alpha)$ being a counterpart of function (2.9.1) possesses the standard properties listed in Theorem 2.9.1 with the replacement of the quantities Ω^*, λ_δ and ν_δ by $\widehat{\Omega}^*$, $\hat{\lambda}_\varkappa$ and $\hat{\nu}$, respectively. On the same grounds as in Section 2.9 in Chapter 2 we will assume $\Omega^* \geq 0$ and, consequently, $\widehat{\Omega}_N^* \geq 0$.

A proper choice of the regularization parameter α_ζ in the **f.d.g.p.s.f.** algorithm amounts to finding a solution of the equation with a monotone function

$$\hat{\varepsilon}(\alpha) = 0 \tag{2}$$

Following Theorems 2.9.2–2.9.4 one can establish Theorems 3 and 4 formulated below.

Theorem 3 *If for any fixed ζ the inequalities $\hat{\varepsilon}_0 \equiv \hat{\varepsilon}(+0) < 0$ and $\hat{\varepsilon}_\infty \equiv \hat{\varepsilon}(+\infty) > 0$ hold, then equation (2) has a unique solution $\alpha_\zeta > 0$. This solution appears to be ordinary when problem (1.5) of minimizing a smoothing functional possesses a unique solution $\hat{z}^\alpha$ for any $\alpha > 0$.*

Theorem 4 *Let the conditions of Theorem 2.2 hold. Then either of the inequalities $\hat{\varepsilon}_0 < 0$ and $\hat{\varepsilon}_\infty > 0$ is valid for all sufficiently small $\|\eta\|$ and $\varkappa$.*

Proof. The proof is carried out as in Theorem 2.2 by merely replacing the quantities $\hat{\rho}_0$ and $\hat{\rho}_\infty$ by $\hat{\varepsilon}_0$ and $\hat{\varepsilon}_\infty$, and inserting the function $\Psi^p(\eta, \Omega)$ in place of $\Psi(\eta, \Omega)$ (see also Theorem 2.9.3). □

Assuming that the regularization parameter $\alpha_\zeta > 0$ is a solution to equation (2), let us introduce the selection rule of the **f.d.g.p.s.f.** algorithm when treating $z_\zeta \equiv \bar{P}_N \hat{z}^{\alpha_\zeta}$ as a solution of the fundamental finite-dimensional problem under consideration.

Selection rule *We choose an element $\hat{z}^{\alpha_\zeta}$ from the set $\widehat{Z}^{\alpha_\zeta}$ in such a way to satisfy the inequality $\widehat{E}(\hat{z}^{\alpha_\zeta}) \geq 0$. For example, one can take $\hat{z}^{\alpha_\zeta} = \hat{z}_+^{\alpha_\zeta}$.*

Theorem 5 *Let, in addition to the conditions of the problem statement from Section 1, $Z^* \cap Z_0 = \varnothing$. Then $\alpha_\zeta \to 0$ as $\zeta \to 0$.*

Proof. The proof is similar to that of Theorem 2.9.5 when α_δ and z^α are replaced by α_ζ and $\bar{P}\hat{z}^\alpha$, respectively. Moreover, the quantities λ_δ, Ω^*, $\Psi(\delta, \Omega)$ have to be formally corrected by their counterparts $\hat{\lambda}_\varkappa$, $\widehat{\Omega}^*$ and $\Psi(\eta, \Omega)$. □

In just the same way as in Theorem 2.9.6, we will prove the following theorem by changing the corresponding quantities and inserting $\bar{\Omega}_\eta \equiv \max\{\bar{\Omega}, \Omega(\bar{P}P\bar{z})\}$ instead of $\bar{\Omega}_\delta$.

Theorem 6 *Let the conditions of Theorem 2.2 hold and the quantities $z_\zeta \equiv \bar{P}\hat{z}^{\alpha_\zeta}$ be obtained by means of the* **f.d.g.p.s.f.** *algorithm. Then for any sequence $\{\zeta_n\}$ converging to 0 as $n \to \infty$ the appropriate sequence $\{z_n\}$, $z_n \equiv z_{\zeta_n}$, satisfies both relations $z_n \xrightarrow{\tau} \bar{Z}$ and $\Omega(z_n) \to \bar{\Omega}$ as $n \to \infty$.*

From this theorem we can draw the conclusion that the **f.d.g.p.s.f.** algorithm also produces a solution of the fundamental finite-dimensional problem.

4.4 Finite-dimensional algorithms for operator equations

In Sections 2–3 we developed the algorithms for solving ill-posed extremal problems on the basis on finite-dimensional data, whose use permits us to construct the appropriate solution. Such theory finds a wide range of applications and, in particular, can be employed to solve operator equations by variational methods. In a common setting of the problem, we are concerned with a topological Hausdorff space (Z, τ) and a metric space U equipped with metric ρ. Furthermore, an operator $A\colon D \to U$ (usually nonlinear) is defined on a set D ($D \subset Z$). Given $(A, \bar{u})$ with $\bar{u} \in U$, it is required to obtain elements $z \in D$ for which

$$Az = \bar{u} \tag{1}$$

Assuming equation (1) to possess a nonempty set of quasisolutions Z^* on D (see Section 3.1 in Chapter 3) we begin by defining the functional $\Omega(z)$, $\Omega(z) \geq \Omega^*$, on the set D and by considering the problem of finding Ω-optimal quasisolutions to equation (1) on the set D in which it is necessary to determine elements $\bar{z} \in Z^*$ such that

$$\Omega(\bar{z}) = \inf\{\Omega(z)\colon z \in Z^*\} \tag{2}$$

The set of solutions of the problem (2) is denoted, as usual, by $\bar{Z}$.

In Section 3.1 in Chapter 3, we investigated the fundamental problem of constructing approximations to the set $\bar{Z}$ from available approximate data of the problem by means of approximations (A_h, u_σ) rather than of the exact data $(A, \bar{u})$. Here an operator A_h from D into U satisfies the condition of approximation

$$\rho(Az, A_h z) \le \psi(h, \Omega(z)) \qquad \forall z \in D$$

and an element $u_\sigma \in U$ is chosen subject to the condition $\rho(\bar{u}, u_\sigma) \le \sigma$. The numbers h, σ and the function ψ are presupposed to be known in advance.

In this section, the quantities (A_h, u_σ) and $(\bar{A}, \bar{u})$ are assumed to be unknown and only their finite-dimensional approximations are available. We focus the reader's attention on the statement of the problem of finite-dimensional approximation in prescribed finite-dimensional normed spaces Z_N and U_M of the dimensions N and M, respectively (see Leonov (1988b,c)). The symbols $\hat{z}$ and $\hat{u}$ denote the points in those spaces: $\hat{z} \in Z_N$ and $\hat{u} \in U_M$. In the framework of Section 1 we refer to operators $P_N\colon Z \to Z_N$ and $\bar{P}_N\colon Z_N \to Z$ possessing all the properties listed in Section 1 and consider a nonempty set $\widehat{D} = \{\hat{z}\colon \bar{P}\hat{z} \in D\}$, which is a finite-dimensional counterpart of the set D. Another continuous operator $\bar{Q}_M\colon U_M \to U$ will complement our studies. Assume that, instead of the approximate data (A_h, u_σ) of the fundamental problem, we have at our disposal only their finite-dimensional approximations including an operator $\widehat{A}_H$, $H \equiv (h, 1/M, 1/N)$, from $\widehat{D}$ into U_M and an element $\hat{u}_\sigma \equiv \hat{u}_{\sigma M} \in U_M$. Let the conditions of finite-dimensional approximation

$$\rho(u_\sigma, \bar{Q}_M \hat{u}_\sigma) \le \xi(1/M, \sigma) \tag{3}$$

$$\rho(A_h z, \bar{Q}_M \widehat{A}_H P_N z) \le \hat{\psi}(H, \Omega(z)) \qquad \forall z \in D \tag{4}$$

hold with $\xi(1/M, \sigma)$ and $\hat{\psi}(H, \Omega(z))$ being the known measures of approximation.

The fundamental finite-dimensional problem of this section consists of processing the approximate data

$$\{\widehat{A}_H, \hat{u}_\sigma, \hat{\psi}, \xi, \psi, h, \delta, M, N, Z_N, U_M\}$$

and constructing an element $\hat{z}_\eta \in \widehat{D}$ $(\eta = (H, \sigma))$ such that the family $\{z_\eta\}$, $z_\eta \equiv \bar{P}_N \hat{z}_\eta$, will converge τ-sequentially to $\bar{Z}$ as $\eta \to 0$

Certain functionals may be of help in achieving this aim. In subsequent studies we deal with

$$J(z) \equiv \rho(Az, \bar{u}) \qquad J_\delta(z) \equiv \rho(A_h z, u_\sigma)$$

$$\Psi_0(\delta, \Omega(z)) \equiv \sigma + \psi(h, \Omega(z)) \qquad z \in D, \quad \delta \equiv (h, \sigma)$$

$$\widehat{J}_\eta(\hat{z}) \equiv \rho(\bar{Q}_M \widehat{A}_H \hat{z}, \bar{Q}_M \hat{u}_\sigma) \qquad \widehat{I}(\hat{z}) \equiv f[\widehat{J}_\eta(\hat{z})]$$

$$\widehat{\Omega}(\hat{z}) \equiv \Omega(\bar{P}\hat{z}) \qquad \widehat{\Psi}(\eta, \Omega(z)) \equiv \xi(1/M, \sigma) + \hat{\psi}(H, \Omega(z))$$

$$\Psi(\eta, \Omega) \equiv \Psi_0(\delta, \Omega) + \widehat{\Psi}(\eta, \Omega) = \sigma + \xi(1/M, \sigma) + \psi(h, \Omega) + \hat{\psi}(H, \Omega)$$

It turns out that the fundamental finite-dimensional problem in the current setting is a particular case of rather general fundamental finite-dimensional problem for ill-posed extremal problems of Section 1.

In addition to the requirements just formulated in the problem statement, Assumptions 1–4 of Section 3.1 in Chapter 3 and properties (1)–(4) of the operators P and $\bar{P}$ from Section 1 are accompanied by the following ones:

(a) $\widehat{A}_H$ are continuous operators from $\widehat{D}$ into U_M for any fixed H;

(b) for $s, \sigma \geq 0$ the functions $\xi(s,\sigma)$ are continuous with $\xi(0,0) = 0$;

(c) the approximation measure $\hat{\psi}(H,\Omega)$ obeys the standard properties of section 2.2 in Chapter 2.

The fulfilment of the whole collection of these conditions provides a possibility of applying the results of Sections 1–3 to solving the fundamental finite-dimensional problem at hand. In particular, in order to find the approximate solutions $z_\eta = \bar{P}_N \hat{z}_\eta$ we need the finite-dimensional generalized principles of discrepancy, quasisolutions and smoothing functional given in Sections 2–3.

Along with the general case, we consider a special case of problem (1) when the operator equation is solvable on D. As in Section 3.3 in Chapter 3, we are working with modifications of finite-dimensional algorithms without using either the value of the **generalized finite-dimensional incompatibility measure**

$$\begin{aligned}\hat{\lambda} &= \inf\left\{\widehat{J}_\eta(\hat{z}) + \Psi(\eta, \widehat{\Omega}(z)) \colon \hat{z} \in \widehat{D}\right\} \\ &= \inf\left\{\rho(\widehat{Q}\widehat{A}_H \hat{z}, \hat{u}_\sigma) + \sigma + \xi(1/M, \sigma)\right. \\ &\qquad \left. + \psi(h, \widehat{\Omega}(\hat{z})) + \hat{\psi}(H, \widehat{\Omega}(\hat{z})) \colon \hat{z} \in \widehat{D}\right\}\end{aligned}$$

or its approximate values. In dealing with solvable equation (1) on the set D, that is, with $\mu_0 \equiv J^* = 0$, we arrange below the **algorithm of the modified finite-dimensional g.p.d. (m.f.d.g.p.d.)**. With this aim, we have occasion to use the functions

$$\hat{\hat{\pi}}(\alpha) \equiv Kf[\Psi(\eta, \widehat{\Omega}(\hat{z}^\alpha))] = K\hat{\hat{\Pi}}(\hat{z}^\alpha)$$

$$\hat{\hat{\rho}}(\alpha) \equiv \hat{\beta}(\alpha) - \hat{\hat{\pi}}(\alpha) = \hat{\hat{P}}(\hat{z}^\alpha) \qquad \hat{z}^\alpha \in \widehat{Z}^\alpha \quad K = \text{const} \geq 1$$

In the **m.f.d.g.p.d.** algorithm, a solution of the equation with a monotonically nondecreasing function

$$\hat{\hat{\rho}}(\alpha) = 0 \qquad\qquad \alpha \geq 0 \tag{5}$$

will be adopted as the regularization parameter α_η.

The existence of such a solution α_η and the conditions for its positiveness have been established in Theorems 2.1–2.2.

The element $\hat{z}_\eta \equiv \hat{z}^{\alpha_\eta}$ involved in the solution of the fundamental finite-dimensional problem of the present section is obtained by the following selection rule of the **m.f.d.g.p.d.** algorithm. As in Section 2, we introduce for given constants C, $q > 1$ the needed values $\alpha_1 = \alpha_\eta/q$, $\alpha_2 = \alpha_\eta q$, $\hat{z}^{\alpha_1}$ and $\hat{z}^{\alpha_2}$.

Selection rule

a) Let $\alpha_\eta > 0$. If the inequality

$$\widehat{I}(\hat{z}^{\alpha_2}) \geq C\hat{\widehat{\Pi}}(\hat{z}^{\alpha_1}) \tag{6}$$

holds, we choose from the set $\widehat{Z}^{\alpha_\eta}$ an element $\hat{z}^{\alpha_\eta}$ for which $\hat{\widehat{P}}(\hat{z}^{\alpha_\eta}) \leq 0$. For example, we might agree with $\hat{z}^{\alpha_\eta} = \hat{z}_-^{\alpha_\eta}$. But if

$$\widehat{I}(\hat{z}^{\alpha_2}) \leq C\hat{\widehat{\Pi}}(\hat{z}^{\alpha_1}) \tag{7}$$

we take an element $\hat{z}^{\alpha_\eta} \in \widehat{Z}^{\alpha_\eta}$ such that $\hat{\widehat{P}}(\hat{z}^{\alpha_\eta}) \geq 0$. In particular, the choice $\hat{z}^{\alpha_\eta} = \hat{z}_+^{\alpha_\eta}$ suits us perfectly.

b) For $\alpha_\eta = 0$ any solution of the extremal problem in which it is required to obtain $\hat{z}_\eta \in \widehat{D}$ such that

$$\widehat{J}_\eta(\hat{z}_\eta) = \inf\left\{\widehat{J}_\eta(\hat{z}) : \hat{z} \in \widehat{Z}_0\right\} \tag{8}$$

can be adopted as the needed element $\hat{z}_\eta$.

As stated in Section 2, problem (8) is solvable.

In this regard, the question of convergence of the approximations $z_\eta = \bar{P}_N \hat{z}_\eta$ will be the subject of special investigations.

Theorem 1 *Let $\{\eta_n\}$ be an arbitrary sequence converging to 0 as $n \to \infty$ and $\alpha_{\eta_n} > 0$ for each n. Let, in addition to the above conditions, the function $f(x)$ defined for $x \geq 0$ be continuous and monotonically increasing. Then the sequence $\{z_n\}$, $z_n \equiv \bar{P}_{N_n} \hat{z}^{\alpha_{\eta_n}}$, constructed by means of the* **m.f.d.g.p.d.** *algorithm satisfies the limit relations*

$$z_n \xrightarrow{T} \bar{Z} \text{ and } \Omega(z_n) \to \bar{\Omega} \text{ as } n \to \infty$$

Proof. The proof is carried out in just the same way as in Theorem 3.3.1 and is based on analogues of Lemmas 3.3.1–3.3.3. The only difference lies in the notations of quantities in the corresponding formulae. In particular, in the analogues of formulae (3.3.8)–(3.3.13) the number

$$\bar{\Omega}_\eta \equiv \max\{\bar{\Omega}, \Omega(\bar{P}_N P_N \bar{z})\}$$

will stand in place of the quantity $\bar{\Omega}$. □

Theorem 2 *Let, in addition to the conditions of the problem statement, the total measure of approximation $\Psi(\eta, \Omega)$ increase in the second argument for any fixed nonzero η. If $\alpha_{\eta_n} = 0$ for a sequence $\{\eta_n\}$, $\eta_n \neq 0$, converging to 0 as $n \to \infty$ and elements $\hat{z}_n \equiv \hat{z}_{\eta_n}$ solve problem (8), then the sequence $\{z_n\}$, $z_n \equiv \bar{P}_{N_n} \hat{z}_n$, satisfies the limit relations*

$$z_n \xrightarrow{T} \bar{Z} \text{ and } \Omega(z_n) \to \bar{\Omega} \text{ as } n \to \infty$$

Proof. The proof is carried out in just the same way as in Theorem 2.10.3. For this reason we point out only some features of current proof. First the inclusion $\hat{z}_\eta \in \widehat{Z}_0$ implies

$$\Omega(\bar{P}\hat{z}_\eta) = \widehat{\Omega}(\hat{z}_\eta) = \widehat{\Omega}^* \leq \Omega(\bar{P}_N P_N \bar{z})$$

Whence, by virtue of property 4) of the operators P and $\bar{P}$, it follows that

$$\overline{\lim}_{\eta\to 0}\Omega(\bar{P}_N\hat{z}_\eta) \leq \overline{\lim}_{\eta\to 0}\Omega(\bar{P}_N P_N \bar{z}) \leq \bar{\Omega} \tag{9}$$

The preceding limit relations constitute what is called an analogue of estimate (2.10.10). Furthermore, joint use of inequalities (1.7)–(1.8), the analogue of Theorem 2.10.1 and the equality $J^* = 0$ gives us the estimate

$$\begin{aligned} 0 - \Psi(\eta, \widehat{\Omega}^*) &\leq \widehat{J}_\eta(\hat{z}_\eta) \equiv J_\eta(\bar{P}\hat{z}_\eta) \\ &= \inf\{\widehat{J}_\eta(\hat{z})\colon \hat{z} \in \widehat{Z}_0\} \\ &= \inf\{\widehat{J}_\eta(\hat{z})\colon \hat{z} \in \widehat{D}\} \\ &\leq \widehat{J}_\eta(\hat{\bar{z}}) \leq 0 + \Psi(\eta, \bar{\Omega}) \end{aligned}$$

which, by analogy with Theorem 2.10.3, implies

$$\lim_{\eta\to 0} J_\eta(\bar{P}_N\hat{z}_\eta) = 0 \tag{10}$$

Due to Lemma 2.5.1, relations (9) and (10) provide the desired convergences. □

Thus, the **m.f.d.g.p.d.** algorithm is best suited to solve the fundamental finite-dimensional problem for the compatible equation (1). We now turn to the **modified f.d.g.p.q.** known as the **m.f.d.g.p.q.** algorithm.

Here, by a proper choice of the regularization parameter α_η, we mean solving equation (3.1) and selecting an element $\hat{z}_n = \hat{z}^{\alpha_\eta}$ by an approved rule governing what can happen.

Selection rule

a) *Let $\alpha_\eta > 0$. If inequality (6) holds, we choose from the set $\widehat{Z}^{\alpha_\eta}$ an element $\hat{z}^{\alpha_\eta}$ in such a way to satisfy the inequality $\widehat{\Omega}(\hat{z}^{\alpha_\eta}) \geq \hat{\omega}_\eta$. Under condition (7) any element $\hat{z}^{\alpha_\eta}$ for which $\widehat{\Omega}(\hat{z}^{\alpha_\eta}) \leq \hat{\omega}_\eta$ can be taken.*

b) *If $\alpha_\eta = 0$, we adopt as $\hat{z}^{\alpha_\eta}$ any solution of the extremal problem in which it is necessary to obtain elements $\hat{z}^0_\eta$ such that*

$$\widehat{J}_\eta(\hat{z}^0_\eta) = \inf\{\widehat{J}_\eta(\hat{z})\colon \hat{z} \in D\} \tag{11}$$

The solvability of problem (11) can be established in a similar way as was done for its analogue (2.10.5) in Theorem 2.10.2. The inequality $\widehat{\Omega}(\hat{z}^0_\eta) \leq \hat{\omega}_\eta$ is still valid in this case.

The following theorem on convergence of the approximations $z_\eta = \bar{P}\hat{z}_\eta$ is valid.

Theorem 3 *Let in the general setting the numbers $\hat{\omega}_\eta$ be such that $\hat{\omega}_\eta \geq \bar{\Omega}$ and $\hat{\omega}_\eta \to \bar{\Omega}$ as $\eta \to 0$. If the function $f(x)$ defined for $x \geq 0$ is continuous and monotonically increasing then, for any sequence $\{\eta_n\}$ tending to 0 as $n \to \infty$, the corresponding sequence $\{z_n\}$, $z_n \equiv z_{\eta_n}$, obtained by means of the* **m.f.d.g.p.q.** *algorithm, possesses the convergence properties: $z_n \xrightarrow{T} \bar{Z}$ and $\Omega(z_n) \to \bar{\Omega}$ as $n \to \infty$.*

Proof. In the case where $\alpha_{\eta_n} > 0$ the proof is similar to that of Theorem 3.2. When $\alpha_{\eta_n} = 0$ for any n relation (11) and the approximation condition (1.7) together imply

$$0 \leq J_\eta(\bar{P}_N \hat{z}_\eta) \leq J_\eta(\bar{P}_N P_N \bar{z}) \leq 0 + \Psi(\eta, \bar{\Omega})$$

Whence it follows that $J_\eta(\bar{P}_N \hat{z}_\eta) \to 0$ as $\eta \to 0$. In light of the property of the problem (11) solutions we arrive at the relations

$$\overline{\lim}_{\eta \to 0} \Omega(\bar{P}_N \hat{z}_\eta) \leq \lim_{\eta \to 0} \hat{\omega}_\eta = \bar{\Omega}$$

To obtain the desired convergences it remains to apply Lemma 2.5.1. □

Finally, we proceed to design the **modified f.d.g.p.s.f.**, the so-called **m.f.d.g.p.s.f.** algorithm, by appeal to the function

$$\hat{\bar{\varepsilon}}(\alpha) \equiv \hat{\varphi}(\alpha) - f[\Psi^p(\eta, \widehat{\Omega}(\hat{z}^\alpha))] \equiv \hat{\bar{E}}(\hat{z}^\alpha)$$

which is a counterpart of the function $\bar{\varepsilon}(\alpha)$ from Section 3.3 in Chapter 3.

The problem of making a substantiated choice of the regularization parameter α_η in the **m.f.d.g.p.s.f.** algorithm is to solve the equation with a monotone function

$$\hat{\bar{\varepsilon}}(\alpha) = 0 \qquad \Omega^* \geq 0 \tag{12}$$

The solution to equation (12) can be found from the result of Theorem 3.3. As in Theorem 2.2, one can prove that for $\|\eta\| \leq \Delta_0$ equation (12) possesses a unique solution $\alpha_\eta > 0$ if $Z_0 \cap Z^* = \varnothing$.

The selection rule of the **m.f.d.g.p.s.f.** algorithm is given below.

Selection rule

a) If $\alpha_\eta > 0$, then an element $\hat{z}_\eta \equiv \hat{z}^{\alpha_\eta}$ is chosen from the set $\widehat{Z}^{\alpha_\eta}$ in such a way to satisfy $\bar{E}(\hat{z}^{\alpha_\eta}) \geq 0$; for example, we can take $\hat{z}^{\alpha_\eta} = \hat{z}_+^{\alpha_\eta}$.

b) If $\alpha_\eta = 0$, then any solution of problem (8) can be adopted as the needed element $\hat{z}_\eta$.

In this way, $z_\eta = \bar{P}_N \hat{z}_\eta$ can be declared to be a solution of the fundamental finite-dimensional problem at hand.

Using similar techniques as in proving Theorems 2.9.2–2.9.4 we will prove the following propositions taking into account the remarks of Theorems 3.4–3.6 and Theorem 2.

Theorem 4 *Let the conditions in the general setting be accompanied by the following one:* $Z^* \cap Z_0 = \varnothing$. *Then the regularization parameter obtained by means of the* **m.f.d.g.p.s.f.** *algorithm satisfies the limit relation* $\alpha_\eta \to 0$ *as* $\eta \to 0$.

Theorem 5 *Let the conditions of the problem statement hold. If* $Z^* \cap Z_0 = \varnothing$ *and the total measure of approximation* $\Psi(\eta, \Omega)$ *increases in the second argument for each nonzero* η, *then for any sequence* $\{\eta_n\}$ *converging to* 0 *as* $n \to \infty$ *the corresponding sequence* $\{z_n\}$, $z_n \equiv \bar{P}_{N_n} \hat{z}_{\eta_n}$, *satisfies the limit relations* $z_n \xrightarrow{T} \bar{Z}$ *and* $\Omega(z_n) \to \bar{\Omega}$ *as* $n \to \infty$.

It follows that the modified finite-dimensional principles of discrepancy, quasisolutions and smoothing functional give regularizing algorithms for solving compatible operator equations.

4.5 Examples of problems

Many important practical problems owe a debt to the variational regularization. Several examples of operator equations frequently encountered in real-life situations bring out the indisputable merit of the regularization method and unveil its potential. To find Ω-optimal solutions of such equations one can apply finite-dimensional algorithms of Sections 2–4. In studying these examples especial attention is paid to particular realizations of the general framework of Section 1 for finite-dimensional approximation of ill-posed problems. With this aim, we agree to consider $h = 0$ and $\sigma = 0$ (see Section 4) enabling us to take into account in the algorithms exclusively the error of finite-dimensional approximation with respect to the ingredients of the problem.

Example 1 Suppose that the function $K(x, s, z)$ is defined on the set $\Pi_0 = \{[c, d] \times [a, b] \times \mathbf{R}\}$ and satisfies the following conditions:

(a) $K(x, s, z) \in C^1(\Pi_0)$;

(b) for each $(x, s, z) \in \Pi_0$, $K(x, s, z)$ admits the collection of estimates

$$|K(x,s,z)| \le \xi_0(|z|) \qquad |K'_x(x,s,z)| \le \xi_1(|z|)$$
$$|K'_s(x,s,z)| \le \xi_2(|z|) \qquad |K'_z(x,s,z)| \le \xi_3(|z|)$$

where the known functions $\xi_m(x)$ $(m = 0, 1, 2, 3)$ defined for $x \ge 0$ are continuous and nondecreasing.

Let us define on the set of all functions $z(s) \in C[a, b]$ an operator acting in accordance with the rule

$$A[x, z(s)] \equiv \int_a^b K(x, s, z(s))ds \qquad x \in [c, d] \tag{1}$$

which is strongly continuous on $W_2^1[a,b]$ into $C[a,b]$ (or $L_2[c,d]$). Indeed, by Lagrange's formula and condition (b) with regard to

$$z^* \equiv \max(\|z_1\|_C, \|z_2\|_C)$$

the inequality

$$|K(x,s,z_1(s)) - K(x,s,z_2(s))| \leq \xi_3(z^*)|z_1(s) - z_2(s)|$$

is valid for arbitrary continuous functions $z_1(s)$ and $z_2(s)$. This provides reason enough to establish the continuity of operator (1) acting from the space $C[a,b]$ into $C[c,d]$ in complete agreement with Example 3.4.4. Therefore, due to Theorem 1.4.9 the complete continuity of the embedding operator from the space $W_2^1[a,b]$ into $C[a,b]$ implies that operator (1) is strongly continuous too.

Let now the space Z coinciding with $W_2^1[a,b]$ be endowed with the topology τ of weak convergence. In the space $U = L_2[c,d]$ with the topology t of convergence in norm we restrict ourselves to the study of the operator equation

$$A[x, z(s)] = \bar{u}(x) \qquad x \in [c,d] \tag{2}$$

having a nonempty set of pseudosolutions $Z^* \subset W_2^1[a,b]$ for the right-hand side $\bar{u}(x) \in C^1[c,d] \subset U$.

Let $g(t)$, $t \geq 0$, be a nonnegative continuous increasing function on Z with $g(+\infty) = +\infty$ and $\Omega(z) = g(\|z\|_{W_2^1})$. We learn from Section 3.4 in Chapter 3 that the functional Ω specified in such a way possesses all of the necessary properties listed in Section 2.2 in Chapter 2. As a corollary to this fact, one can see that the set $\bar{Z} \subset W_2^1[a,b]$ of all normal pseudosolutions to equation (2) appears to be nonempty (see Theorem 2.4.1 and Section 3.1 in Chapter 3). We will assume, in addition, that $\bar{Z} \subset C^1[a,b]$.

With these ingredients, we may set up the problem of approximate determination of normal pseudosolutions to equation (2) by means of finite-dimensional approximations of $(A, \bar{u})$. We expound certain exploratory devices for obtaining them by introducing, for the sake of simplicity, two equidistant grids on intervals $[a,b]$ and $[c,d]$:

$$\begin{aligned} \omega_N(s) &= \{s_j\colon s_j = a + h_s(j-1), j = 1,\dots,N\} \quad h_s = (b-a)/(N-1) \\ \omega_M(x) &= \{x_i\colon x_i = c + h_x(i-1), i = 1,\dots,M\} \quad h_x = (d-c)/(M-1) \end{aligned}$$

After that, the spaces of all grid functions defined on $\omega_N(s)$ and $\omega_M(x)$, respectively, are treated as Z_N and U_M, that is, $\hat{z} = (z_1,\dots,z_N) \in Z_N$ and $\hat{u} = (u_1,\dots,u_M) \in U_M$. We refer to a simple shift operator on the grid (Gavurin (1971) and Samarskiĭ (1989)) that assigns to every function $z(s) \in W_2^1[a,b] = Z$ a vector $P_N z = (z(s_1),\dots,z(s_N)) \in Z_N$. That operator can be viewed as the needed operator P_N. The operators $\bar{P}_N$ and $\bar{Q}_M$ are continuous filling operators converting the values of grid functions by means of interpolation (Gavurin (1971)). Just for this reason properties (1)–(2)

of the operators P and $\bar{P}$ from Section 1 are satisfied. Concrete examples of operators $\bar{P}_N$ possessing properties (3)–(4) from Section 1 will be given special investigation.

Given a grid function $\hat{\bar{u}} = (\bar{u}(x_1), \dots, \bar{u}(x_M))$ instead of the right-hand side $\bar{u}(x)$ of equation (2), we now consider for $x \in [c,d]$ the simplest operator $\bar{Q}_M \colon U_M \to U$ with the values

$$(\bar{Q}_M \hat{u})(x) = \{\bar{u}(x_i) \colon x_i \le x < x_{i+1}, i = 1, \dots, M-1; \bar{u}(x_M) \colon x = x_M\}$$

As $\bar{u} \in C^1[c,d]$, one can establish a concrete approximation condition of the form (4.3) by direct calculations

$$\begin{aligned}
\rho(\bar{u}, \bar{Q}_M \hat{\bar{u}}) &= \|\bar{u} - \bar{Q}_M \hat{\bar{u}}\|_{L_2[c,d]} = \left\{ \sum_{i=1}^{M-1} \int\limits_{x_i}^{x_{i+1}} \left[\bar{u}(x) - \bar{u}(x_i)\right]^2 dx \right\}^{1/2} \\
&\le \left\{ \sum_{i=1}^{M-1} \int\limits_{x_i}^{x_{i+1}} \|\bar{u}\|_{C^1}^2 (x - x_i)^2 dx \right\}^{1/2} \le \|\bar{u}\|_{C^1} \left\{ \sum_{i=1}^{M-1} \frac{h_x^3}{3} \right\}^{1/2} \\
&= h_x \|\bar{u}\|_{C^1} \sqrt{\tfrac{1}{3}(d-c)} \equiv \xi(M^{-1}, 0) \equiv \xi(M^{-1}) \qquad (3)
\end{aligned}$$

which characterizes the property that the function $\xi(M^{-1})$ is the measure of finite-dimensional approximation of the right-hand side $\bar{u}(x)$ by the grid function $\hat{\bar{u}}$.

Furthermore, we shall need a continuous finite-dimensional operator

$$\hat{A}_H[x_i, \hat{z}] \equiv \hat{A}_H \hat{z} = \sum_{j=1}^{N} K(x_i, s_j, z_j) h_s$$

which is defined for $x_i \in \omega_M(x)$, acts into the space U_M from Z_N and represents the simplest finite-dimensional approximation of operator (1). We may attempt a proper approximation condition for the finite-dimensional operator $\hat{A}_H$ in the form (1.4).

According to the definition of the operator $\bar{Q}_M$ we have

$$\begin{aligned}
\rho^2(A[x, z(s)], \bar{Q}_M \hat{A}_H P_N z) &= \|A[x, z(s)] - \bar{Q}_M \hat{A}_H P_N z\|^2_{L_2[c,d]} \\
&= \sum_{i=1}^{M-1} \int\limits_{x_i}^{x_{i+1}} \Bigg\{ \int\limits_a^b K(x, s, z(s)) ds \\
&\quad - \sum_{j=1}^{N} h_s K(x_i, s_j, z(s_j)) \Bigg\}^2 dx \\
&\equiv \sum_{i=1}^{M-1} \int\limits_{x_i}^{x_{i+1}} Q_i^2(x) dx \qquad (4)
\end{aligned}$$

From such reasoning it seems clear that the variables $Q_i(x)$ admit for $x \in [x_i, x_{i+1}]$ $(i = 1, \ldots, M-1)$ the composite estimates

$$|Q_i(x)| \le \sum_{j=1}^{N-1} \int_{s_j}^{s_{j+1}} |K(x,s,z(s)) - K(x_i,s_j,z(s_j))|ds + |K(x_i,s_N,z(s_N))|h_s \equiv Q_{i1} + Q_{i2} \tag{5}$$

The quantity Q_{i2} can be most readily evaluated. In light of property (b) of the function K we arrive at the relations

$$Q_{i2} \le \|K\|_{C(\Pi_0)} h_s \le \xi_0(\|z\|_C) h_s \le h_s \xi_0(k\|z\|_{W_2^1}) \tag{6}$$

Here a constant $k > 0$ is taken from the well-known estimate

$$\|z\|_{C[a,b]} \le k\|z\|_{W_2^1[a,b]}$$

which is valid for any function $z(s) \in W_2^1[a,b]$ (see Trenogin (1980, p. 104) and Section 1.4 in Chapter 1).

In the estimation of Q_{i1} certain properties of the function K prove to be useful and allow the following manipulations

$$\begin{aligned} |K(x,s,z(s)) - K(x_i,s_j,z(s_j))| &\le \|K'_x\|_{C(\Pi_0)}(x - x_i) \\ &\quad + \|K'_s\|_{C(\Pi_0)}(s - s_j) \\ &\quad + \|K'_z\|_{C(\Pi_0)}|z(s) - z(s_j)| \\ &\le \xi_1(k\|z\|_{W_2^1})h_x \\ &\quad + \xi_2(k\|z\|_{W_2^1})(s - s_j) \\ &\quad + \xi_3(k\|z\|_{W_2^1})|z(s) - z(s_j)| \end{aligned}$$

for all $x \in [x_i, x_{i+1}]$ and $s \in [s_j, s_{j+1}]$. Then, applying the preceding result yields

$$\begin{aligned} Q_{i1} &\le \xi_1(k\|z\|_{W_2^1})h_x(b-a) + \xi_2(k\|z\|_{W_2^1}) \\ &\quad \times \sum_{j=1}^{N-1} \int_{s_j}^{s_{j+1}} (s - s_j)ds \\ &\quad + \xi_3(k\|z\|_{W_2^1}) \sum_{j=1}^{N-1} \int_{s_j}^{s_{j+1}} |z(s) - z(s_j)|ds \\ &= (b-a)\Big[\xi_1(k\|z\|_{W_2^1})h_x + \tfrac{1}{2}\xi_2(k\|z\|_{W_2^1})h_s\Big] \\ &\quad + \xi_3(k\|z\|_{W_2^1}) \sum_{j=1}^{N-1} \int_{s_j}^{s_{j+1}} |z(s) - z(s_j)|ds \end{aligned} \tag{7}$$

It remains to evaluate only the last term in (7). Under the condition $z(s) \in W_2^1[a,b]$ the function $z(s)$ is absolutely continuous. Putting certain properties of absolutely continuous functions from Theorem 1.4.10 together with the Cauchy–Bunyakovskii inequality, we are led to the estimate

$$\begin{aligned}
\sum_{j=1}^{N-1} \int_{s_j}^{s_{j+1}} |z(s) - z(s_j)| ds &= \sum_{j=1}^{N-1} \int_{s_j}^{s_{j+1}} \left| \int_{s_j}^{s} z'(t) dt \right| ds \\
&\le \sum_{j=1}^{N-1} \int_{s_j}^{s_{j+1}} \left[\int_{s_j}^{s} |z'(t)| dt \right] ds \\
&\le \sum_{j=1}^{N-1} h_s \int_{s_j}^{s_{j+1}} |z'(t)| dt \\
&= h_s \int_a^b |z'(t)| dt \le h_s \sqrt{b-a} \\
&\quad \times \left\{ \int_a^b |z'(t)|^2 dt \right\}^{1/2} \\
&\le h_s \sqrt{b-a} \|z\|_{W_2^1} \qquad (8)
\end{aligned}$$

Therefore, (7) and (8) provide the validity of the inequality

$$\begin{aligned}
Q_{i1} &\le (b-a)\left[\xi_1(k\|z\|_{W_2^1})h_x + \tfrac{1}{2}\xi_2(k\|z\|_{W_2^1})h_s\right] \\
&\quad + \sqrt{b-a}\|z\|_{W_2^1}\xi_3(k\|z\|_{W_2^1})h_s
\end{aligned}$$

Relations (4)–(6) along with the last inequality imply

$$\begin{aligned}
&\|A[x, z(s)] - \bar{Q}_M \hat{A}_H P_N z\|_{L_2} \\
&\le \sqrt{d-c}\Big\{ \big[\xi_0(k\|z\|_{W_2^1}) \\
&\quad + \frac{1}{2}\xi_2(k\|z\|_{W_2^1})(b-a) \\
&\quad + \sqrt{b-a}\|z\|_{W_2^1}\xi_3(k\|z\|_{W_2^1})\big] h_s \\
&\quad + (b-a) h_x \xi_1(k\|z\|_{W_2^1}) \Big\}
\end{aligned}$$

In complete agreement with (1.4) this provides support for the view that the function

$$\begin{aligned}\widehat{\psi}(H,\Omega) &\equiv \sqrt{d-c}\{[\xi_0[kg^{-1}(\Omega)] + \frac{1}{2}\xi_2[kg^{-1}(\Omega)](b-a) \\ &+\sqrt{b-a}g^{-1}(\Omega)\xi_3[kg^{-1}(\Omega)]]h_s \\ &+(b-a)\xi_1[kg^{-1}(\Omega)]h_x\} \\ H &\equiv (h_x, h_s) \end{aligned} \tag{9}$$

is just the measure of finite-dimensional approximation of the problem at hand. By virtue of the properties of the functions g and ξ_m $(m = 0, 1, 2, 3)$ so defined it is easily verified that the measure $\widehat{\psi}$ of finite-dimensional approximation satisfies the standard requirements of Section 2.2 in Chapter 2.

To decide for yourself whether the algorithms of Sections 2–4 are acceptable for solving the operator equation (2), a first step is to refer to an operator $\bar{P}_N \colon Z_N \to Z$. For the sake of definition, let $\bar{P}_N$ be the operator of piecewise linear interpolation of the values of the grid function $\hat{z} \in Z_N$ between the grid nodes:

$$\begin{aligned}(\bar{P}\hat{z})(s) &= \{(z_{j+1} - z_j)(s - s_j)/h_s + z_j, s \in [s_j, s_{j+1}]\} \\ & \quad j = 1, \ldots, N-1\end{aligned}$$

Then property (3) from Section 1 is valid, since $\bar{P}Pz \in W_2^1[a,b]$ for any function $z \in W_2^1[a,b]$. It only remains to check property (4) for P and $\bar{P}$ both.

Lemma 1 *For any function* $z(s) \in C^1[a,b]$

$$\lim_{N\to\infty} \|\bar{P}_N P_N z\|_{W_2^1[a,b]} = \|z\|_{W_2^1[a,b]}$$

Proof. First of all, it is worth noting one thing. For any element, $\hat{z} = (z_1, \ldots, z_N) \in Z_N$ we have

$$\begin{aligned}\|\bar{P}_N\hat{z}\|^2_{W_2^1} &= \sum_{j=1}^{N-1} \int_{s_j}^{s_{j+1}} \left\{\frac{(z_{j+1} - z_j)(s - s_j)}{h_s} + z_j\right\}^2 ds \\ &+ \sum_{j=1}^{N-1} \int_{s_j}^{s_{j+1}} \left(\frac{z_{j+1} - z_j}{h_s}\right)^2 ds \\ &= \sum_{j=1}^{N-1} z_j^2 h_s + \sum_{j=1}^{N-1} \left[\frac{(z_{j+1} - z_j)^2}{3} + z_j(z_{j+1} - z_j)\right] h_s \\ &+ \sum_{j=1}^{N-1} \left(\frac{z_{j+1} - z_j}{h_s}\right)^2 h_s\end{aligned}$$

The resulting expression with respect to $\hat{z} = P_N z$ for $z \in C^1[a,b]$ is of alternative form

$$\|\bar{P}_N P_N z\|^2_{W_2^1} = \sum_{j=1}^{N-1} z^2(s_j)h_s + h_s \sum_{j=1}^{N-1} \left\{ \tfrac{1}{3}[z'(s_j^*)]^2 h_s^2 + z(s_j)z'(z_j^*)h_s \right\}$$
$$+ \sum_{j=1}^{N-1} [z'(s_j^*)]^2 h_s \equiv r_1(N) + r_2(N) + r_3(N)$$

Here, on the same grounds, we used the Lagrange formula which looks like this

$$z(s_{j+1}) - z(s_j) = z'(s_j^*)h_s \qquad s_j^* \in [s_j, s_{j+1}]$$

The values $r_l(N)$ $(l = 1,2,3)$ will be given special investigation.

With the aid of the relation

$$|r_2(N)| \le \tfrac{1}{3}h_s^2 \|z\|_{C^1}(b-a) + h_s \|z\|_C \|z\|_{C^1}(b-a)$$

which is valid for any $z(s) \in C^1[a,b]$, it is not difficult to establish $r_2(N) \to 0$ as $N \to \infty$ or, what amounts to the same, as $h_s \to 0$. Furthermore, since the functions $z(s)$ and $z'(s)$ are summable on the segment $[a,b]$, the values $r_1(N)$ and $r_3(N)$ converge to integral sums as $h_s \to 0$:

$$\lim_{N\to\infty} r_1(N) = \int_a^b z^2(s)ds \qquad \lim_{N\to\infty} r_3(N) = \int_a^b [z'(s)]^2 ds$$

Eventually, we arrive at

$$\lim_{N\to\infty} \|\bar{P}_N P_N z\|^2_{W_2^1[a,b]} = \int_a^b \{z^2(s) + [z'(s)]^2\} ds = \|z\|^2_{W_2^1}$$

thereby completing the proof of the lemma. □

It is worth noting here that the validity of property (4) from Section 1 is a consequence of the inclusion $\bar{Z} \subset C^1[a,b]$, the form of the functional $\Omega(z)$ and Lemma 1:

$$\lim_{N\to\infty} \Omega(\bar{P}_N P_N \bar{z}) = \lim_{N\to\infty} g(\|\bar{P}_N P_N \bar{z}\|_{W_2^1}) = g(\|\bar{z}\|_{W_2^1}) = \bar{\Omega}$$

In accordance with what has been said we draw the conclusion that all the conditions involved in the general setting of Section 4 are satisfied and the **f.d.g.p.d.**, **f.d.g.p.q.** and **f.d.g.p.f.** algorithms with available data $\{\hat{A}_H, \hat{u}, h_s(N), h_x(M), \xi, \widehat{\psi}\}$ are acceptable for approximating the set $\bar{Z}$ of all normal pseudosolutions to equation (2). It should be noted that the algorithms in this example are based on minimizing a smoothing functional of the special form

$$\widehat{M}^\alpha[z] = \alpha g(\|\bar{P}_N \hat{z}\|_{W_2^1}) + f(\|Q_M \hat{A}_H \hat{z} - Q_M \hat{u}\|_{L_2})$$

over the set Z_N. Due to the general theorems on convergence of approximations from Sections 2–3 the families $\{\bar{P}_N \hat{z}_H\}$, $H = (h_x, h_s)$, consisting of all approximate solutions obtained by means of the algorithms, possess the convergence properties:

$$\bar{P}_N \hat{z}_H \xrightarrow{W_2^1} \bar{Z} \qquad \|\bar{P}_N \hat{z}_H\|_{W_2^1} \to g^{-1}(\bar{\Omega})$$

as $H = (h_x, h_s) \to 0$ (in fact, as $M, N \to \infty$). From the results of Section 3.4 in Chapter 3, it follows that the preceding relations imply the strong convergence of approximations:

$$\bar{P}_N \hat{z}_H \xrightarrow{W_2^1} \bar{Z} \qquad \text{as} \qquad H \to 0$$

Observe that rather complicated quadrature approximations of operator (1) and other types of the operators $\bar{Q}$ and $\bar{P}$ can be treated in a similar way after replacing the coefficients in formulae (3) and (9) which indicate the approximation measures. The finite element method also can be fitted into our approximation scheme.

Example 2 Of special interest is one particular case of equation (2):

$$A[x, z(s)] \equiv \int_a^b K_0(x,s) f_0[s, z(s)]ds = \bar{u}(x) \qquad x \in [c,d] \qquad (10)$$

whose quasisolutions are sought on the set

$$D_+ = \{z(s) \in W_2^1[a,b] \colon z(s) \geq 0\}$$

provided that

$$K_0(x,s) \in C^1(\Pi), \quad \Pi \equiv [c,d] \times [a,b], \quad f_0(s,z) \in C(\Pi_1), \quad \Pi_1 \equiv [a,b] \times \mathbf{R}$$

and the estimates

$$|f(s,z)| \leq \xi(|z|) \qquad |f_s'(s,z)| \leq \xi_1(|z|) \qquad |f_z'(s,z)| \leq \xi_2(|z|)$$

are valid at every point $(s,z) \in \Pi_1$. Here ξ, ξ_1 and ξ_2 are continuous monotonically nondecreasing functions as in Example 1.

Setting $Z \equiv W_2^1[a,b]$ and $U \equiv L_2[c,d]$ we can verify, by exactly the same reasoning as in Example 1, that the operator A in equation (10) and acting from Z into U (and from D_+ into U) is strongly continuous. Here we keep in mind the results of Example 3.4.3.

Suppose that the set Z^* of quasisolutions of the equation (10) on the set D_+ is nonempty and the set $\bar{Z}$ of its normal quasisolutions consists of continuously differentiable functions on $[a,b]$.

By applying the finite-dimensional approximation scheme of Example 1 we are concerned with a finite-dimensional continuous operator acting from Z_N into U_M in accordance with the rule

$$\hat{A}_H[x_i, \hat{z}] = \hat{A}_H \hat{z} = \sum_{j=1}^{N} K_0(x_i, s_j) f_0(s_j, z_j) h_s \qquad x_i \in \omega_M(x)$$

Likewise, the approximation condition is expressed by

$$\|A[x,z(s)] - \bar{Q}_M \hat{A}_H P z\|_{L_2[c,d]} \leq \widehat{\psi}(H,\Omega)$$

with

$$\begin{aligned}\widehat{\psi}(H,\Omega) \quad \equiv \quad & \sqrt{d-c}\Big\{\big[(\|K_0\|_{C(\Pi)} + \tfrac{1}{2}(b-a)\|K'_{0s}\|_{C(\Pi)}) \\ & \times\xi[kg^{-1}(\Omega)] + \tfrac{1}{2}\xi_1[kg^{-1}(\Omega)]\|K_0\|_{C(\Pi)}(b-a) \\ & +\sqrt{b-a}\|K_0\|_{C(\Pi)}g^{-1}(\Omega)\xi_2[kg^{-1}(\Omega)]\big]h_s \\ & +(b-a)\|K'_x\|_{C(\Pi)}\xi[kg^{-1}(\Omega)]h_x\Big\}\end{aligned}$$

where, for the same reason as before, the approximation measure satisfies all the standard requirements. As in Example 1, the remaining conditions under which the algorithms of the generalized finite-dimensional principles are widely used, hold true. In the given example these algorithms yield also strong convergence of approximations to $\bar{Z}$ in $W_2^1[a,b]$.

Example 3 Let a function $K(x,s,\gamma)$ be defined on the set

$$\Pi \equiv \{(x,s)\colon 0 \leq x \leq T, 0 \leq s \leq x\} \times \mathbf{R}_+$$

and possess thereon the following properties:

(a) $K(x,s,\gamma) \in C(\Pi)$;

(b) $K(x,s,\gamma)$ is differentiable in x, s for every fixed $\gamma \geq 0$;

(c) $|K(x,s,\gamma)| \leq \xi_0(\gamma)$, $|K'_x(x,s,\gamma)| \leq \xi_1(\gamma)$ and $|K'_s(x,s,\gamma)| \leq \xi_2(\gamma)$ for any $(x,s,\gamma) \in \Pi$. Here $\xi_m(\gamma)$ $(m = 0,1,2)$ are continuous functions, while the functions $\xi_1(\gamma)$ and $\xi_2(\gamma)$ defined for $\gamma \geq 0$ are increasing with the limiting values $\xi_1(+\infty) = +\infty$ and $\xi_2(+\infty) = +\infty$.

We now raise the question of finite-dimensional approximation in a reflexive Banach space

$$Z = W_2^1[0,T] \times \mathbf{R}$$

with elements $z \equiv (z(s),\gamma)$ and norm $\|z\|^2 \equiv \|z(s)\|^2_{W_2^1[0,T]} + \gamma^2$. Given $D \equiv \{z \in Z\colon z(s) \geq 0, \gamma \geq 0\}$ and $U = L_2[0,T]$, one associates with

$$A[x,z] \equiv A[x;z(s),\gamma] = \int_0^x K(x,s,\gamma)z(s)ds$$

an operator from D into U. For the purposes of the present example we endow Z with a relevant topology τ induced by weak convergence in $W_2^1[0,T]$ and convergence in $\mathbf{R}$. Then the inequality

$$\begin{aligned}&|K(x,s,\gamma_1)z_1(s) - K(x,s,\gamma_2)z_2(s)| \\ &\quad \leq \|z_1\|_C|K(x,s,\gamma_1) - K(x,s,\gamma_2)| + \xi_0(\gamma_2)\|z_1 - z_2\|_C\end{aligned}$$

is valid for arbitrary elements $z_1, z_2 \in Z$. Under the conditions imposed on the functions K and ξ_0 this ensures the continuity of the operator A on $C[0,T] \times \mathbf{R}$ into U and its strong continuity on Z into U too.

We now turn to the equation

$$A[x; z(s), \gamma] = \bar{u}(x) \qquad x \in [0,T] \tag{11}$$

related to the two unknowns: a nonnegative function $z(s) \in W_2^1[0,T]$ and an unknown parameter $\gamma \geq 0$. Suppose that a solution $\bar{z} \equiv (\bar{z}(s), \bar{\gamma})$ to equation (11) corresponds to the right-hand side $\bar{u}(x) \in C^1[0,T]$ of that equation and $\bar{z}(s) \in C^1[0,T]$, $\bar{z}(s) \geq 0$, $\bar{\gamma} \geq 0$, that is, $\bar{z} \in D$.

In order to construct approximations to $\bar{z}$, a reasonable form of the functional is

$$\Omega(z) \equiv \Omega[z(s), \gamma] = \|z\|^2_{W_2^1[0,T]} + p_0\xi_0^2(\gamma) + p_1\xi_1^2(\gamma) + p_2\xi_2^2(\gamma)$$

where constants $p_0, p_1, p_2 > 0$ will be specified below. We claim that the functional so defined possesses the properties listed in Section 2.2 in Chapter 2 and used in Section 4. Indeed, the norm $\|z\|_{W_2^1}$ is weakly lower semicontinuous on a Hilbert space $W_2^1[0,T]$ (see, for example, Section 1.3 in Chapter 1) and the functions $\xi_0(\gamma)$, $\xi_1(\gamma)$, $\xi_2(\gamma)$ are continuous for all $\gamma \geq 0$. In view of this, the functional Ω is τ-sequentially lower semicontinuous on the set D being τ-sequentially closed in Z. Moreover, the inequality $\Omega(z) \leq C^2$ implies that $\|z\|_{W_2^1[0,T]} \leq C^2$ and $0 \leq \gamma \leq \max\{\xi_1^{-1}(C/\sqrt{p_1}), \xi_2^{-1}(C/\sqrt{p_2})\}$. Therefore, due to the weak compactness of the ball in the space $W_2^1[0,T]$ and the compactness of the segment in the space $\mathbf{R}$ the set

$$\Omega_{C^2} \equiv \{z \in D \colon \Omega(z) \leq C^2\}$$

is τ-sequentially compact.

Returning to finite-dimensional approximation of equation (11), we introduce the grid

$$\omega_N(s) = \omega_M(x) = \{s_j \colon s_j = H(j-1), j = 1, \dots, N\}$$
$$H = T/(N-1) \qquad M = N$$

and the corresponding space $Z'_N \equiv U_N$ ($\hat{z} = (z_1, \dots, z_N) \in Z'_N$) of all grid functions, whose use permit us to define the space $Z_N \equiv Z'_N \times \mathbf{R}$ the elements of which are identical with $(\hat{z}, \gamma)$ and the operator P_N in Z acting in accordance with the rule

$$\begin{aligned} P_N z &= (z(s_1), \dots, z(s_N), \gamma) \\ &\quad \forall z = (z(s), \gamma) \in Z \qquad s_1, \dots, s_N \in \omega_N(s) \end{aligned}$$

One more operator $\bar{P}_N \colon Z_N \to Z$ complements our studies and is aimed to assign to every element $(\hat{z}, \gamma) \in Z_N$ the values $(\hat{z}(s), \gamma) \in Z$, where a function $\hat{z}(s) \in W_2^1[0,T]$ is obtained by means of linear interpolation of the values of the grid function $\hat{z} \in Z'_N$ between the grid nodes. For subsequent arguments we refer also to the operator $\bar{Q}_N : U_N \to U$ defined by the relation

$$(\bar{Q}_N\hat{u})(x) = \{u_1 : x \in [x_1, x_2]; u_i : x \in (x_i, x_{i+1}], i = 2, \ldots, N-1\}$$

We may attempt a finite-dimensional operator in the form

$$\hat{A}_H(\hat{z}, \gamma) = \sum_{j=1}^{i} K(x_i, s_j, \gamma) z_j H$$

Clearly, it is continuous on Z_N into U_N. Taking into account the restrictions on the function K and the belonging $z(s) \in W_2^1[0,T]$ and following the guidelines of Example 1 we obtain the estimate

$$\begin{aligned} \|A[x; z(s), \gamma] - \bar{Q}_N \hat{A}_H P_N z\|_{L_2[0,T]} &\le \widehat{\psi}(H, \Omega) \\ \widehat{\psi}(H, \Omega) \quad &\equiv \quad \tfrac{1}{2}\sqrt{T} h \{ \|z(s)\|^2_{W_2^1[0,T]} \\ &\quad + p_0 \xi_0^2(\gamma) + p_1 \xi_1^2(\gamma) \\ &\quad + p_2 \xi_2^2(\gamma) \} = \tfrac{1}{2}\sqrt{T} h \Omega(z) \end{aligned} \tag{12}$$

with $p_0 = 3(\sqrt{T} + k)^2$ and $p_1 = p_2 = 3k^2T^2$. The approximation measure $\widehat{\psi}(H, \Omega)$ is consistent with the assumptions of Section 2.2 in Chapter 2.

The approximation of the right-hand side of equation (11) can be constructed in just the same way as we did in Example 1.

Using exactly the same reasoning as before, one can employ the modified finite-dimensional algorithms of Section 4 for finding solutions $\bar{z} = (\bar{z}(s), \bar{\gamma})$ to equation (11). Thus, using the results of Examples 1–2 we establish the convergence of approximations to $\bar{z}$ in the Z-norm as $H \to 0$.

Equation (11) with the kernel $K(x, s, \gamma) = \exp\{-\gamma(x - s)\}$ $(\gamma \ge 0)$ may be of help in further elaboration on this subject. It is clear that the conditions imposed on the function K are satisfied and $\xi_0(\gamma) = 1$, $\xi_1(\gamma) = \xi_2(\gamma) = \gamma$. In so doing, some quadratic functionals with respect to $z(s)$ and γ can serve as Ω, namely

$$\Omega(z) = \|z(s)\|^2_{W_2^1[0,T]} + 6k^2T^2\gamma^2$$

In each of those cases, the approximation measure involved in (12) becomes

$$\widehat{\psi}(H, \Omega) = \tfrac{1}{2}\sqrt{T} h \{\Omega + p_0\}$$

There are plenty of practical problems for which finite-dimensional generalized principles of discrepancy, quasisolutions and smoothing functional provide an excellent background. Slight modifications of the given examples with other types of approximation, convergence, etc. are widely used and are gaining increasing popularity.

In conclusion we give some remarks on the numerical realization of algorithms from Sections 2–4. The methods of such realizations were discussed in Section 2.14 in Chapter 2. The only difference lies in the utilization of finite-dimensional functionals. On the basis of these methods, a complex of programs has been developed by Leonov (1988 b,c) to solve nonlinear integral equations of the forms (2), (10), (11) and some others.

While solving such problems, it is fairly common to adopt a combination of the Fletcher–Reeves method with some modifications and (see Polak (1971)) the Newton method as the basic technique of minimizing a smoothing functional and the method for the approximate determination of the generalized incompatibility measure. We recommend the readers to such a combined method since it possesses certain advantages concerning the rate of convergence.

To demonstrate our approach, we consider several examples of solving some modeling problems, using the complex of programs mentioned above. Common practice in the theory of nonlinear ill-posed problems involves an integral equation of the form (2)

$$\int_a^b \log \frac{(x-\xi)^2 + H^2}{(x-\xi)^2 + (\bar{z}(\xi) - H)^2} d\xi = \bar{u}(x) \qquad x \in [c, d] \tag{13}$$

which arose from some problems of geophysics (see Tikhonov and Glasko (1965)) and may be interpreted in our testing as the standard model, where the parameter $H > 0$ is known in advance and the desired solution is exactly $\bar{z}(s)$. Under additional assumptions $\bar{z}(\xi) \in W_2^1[a,b]$ and $\bar{z}(\xi) \geq H + \varepsilon$, where $\varepsilon > 0$ is a given number, the resulting equation for $z(\xi)$ obtained from (13) by substituting $z^2(\xi) = \bar{z}(\xi) - H - \varepsilon$ can be approximated by the same scheme as in Example 1 with $M = N = 41$, $a = 0$, $b = 1$, $c = -1$ and $d = 2$. Here the function $\bar{z}(\xi) = \xi(1-\xi) + H + \varepsilon$ defined for $0 \leq \xi \leq 1$ is treated as a model solution. The function $\bar{u}(x)$ for $x \in [-1, 2]$ is recovered from (13) via the function $\bar{z}(\xi)$. Then $\bar{u}(x)$ is perturbed by a noise of the form $\sigma\omega$, where ω is a random variable with the uniform distribution on $[-1, 1]$. Here a reasonable form of the functional Ω is $\Omega(z) = \|z\|^2_{W_2^1[a,b]}$. In applications of the algorithms from Section 4 we keep $f(x) = x^2$, $C = 1.01$, $q = 1.01$, $\omega_\eta = 0.4$, $p = 9/10$ and $\varepsilon = 10^{-5}$ and take a constant function $\bar{z}_0(\xi) = 1 + H + \varepsilon$ as the initial approximation of the minimization method in question.

The results of solving equation (13) by means of the **f.d.g.p.d.**, **f.d.g.p.q.** and **f.d.g.p.f.** algorithms, respectively, are given in Fig. 4.1 at various perturbation levels σ of the right-hand side $\bar{u}$ (the exact solution $\bar{z}$ is plotted by continuous curves – and the approximate solutions are depicted by symbols ∘∘∘ for $\sigma = 10^{-4}$ and by symbols ▲▲▲ for $\sigma = 10^{-3}$).

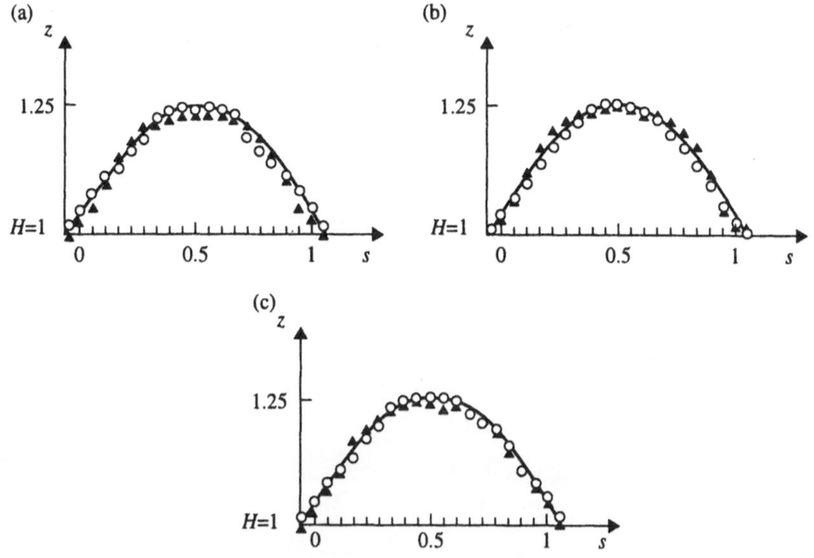

Figure 4.1 Applying of the algorithms of the generalized principles to the integral equation (13): (a) **f.d.g.p.d.**; (b) **f.d.g.p.q.**; (c) **f.d.g.p.f.**

CHAPTER 5

Piece-uniform regularization of ill-posed problems with discontinuous solutions

When solving ill-posed problems by regularization methods one can encounter the situations in which it is necessary to produce piece-uniform approximations to an exact solution having, for example, a finite number of discontinuity points of the first kind with unknown positions. In this chapter we show how to resolve these problems by using functions of bounded variation (see Leonov (1980, 1982a, 1987b)).

5.1 Some properties of bounded variation functions

Denote by $V[a,b]$ the space of bounded variation functions defined on the segment $[a,b]$. We obtain the Banach space $V[a,b]$ when the norm on it is defined by

$$\|z\| \equiv \|z\|_{V[a,b]} = |z(a)| + \bigvee_a^b(z) \qquad \forall z \in V[a,b]$$

All the functions of bounded variation possess two special properties: they are bounded and have at most countable set of discontinuity points of the first kind (see Section 1.4 in Chapter 1).

As before, it is convenient to introduce several notations. Let $s_1, \ldots, s_N$ be points of the segment $[a,b]$ subject to the condition $a = s_1 < \cdots < s_N = b$. Any such collection $\Pi_N = \{s_1, \ldots, s_N\}$ is called a partition of the segment $[a,b]$. For a bounded function $z(s)$ on $[a,b]$ and a partition Π_N set

$$\omega(z, \Pi_N) = \sum_{k=1}^{N-1} \left|z(s_{k+1}) - z(s_k)\right|$$

Observe that for any $z(s) \in V[a,b]$ we might have

$$\omega(z, \Pi_N) \leq \bigvee_a^b(z) \tag{1}$$

disregarding to the partition Π_N (see Natanson (1974)).

In the sequel we will use the results concerning convergence of sequences of bounded variation functions.

Lemma 1 *Let a sequence of functions $\{z_n(s)\}_{n=1}^{\infty} \subset V[a,b]$ possess the following properties:*

(a) *$z_n(s)$ converges to a function $\bar{z}(s)$ as $n \to \infty$ for each $s \in [a,b]$;*

(b) *$\|z_n\| \leq \bar{M} = const$ for each n.*

Then $\bar{z}(s) \in V[a,b]$ and

$$\bigvee_a^b(\bar{z}) \leq \underline{\lim}_{n\to\infty} \bigvee_a^b(z_n) \qquad \|\bar{z}\| \leq \underline{\lim}_{n\to\infty}\|z_n\|$$

Proof. Consider an arbitrary partition Π_N of the segment $[a,b]$ to be fixed. Then the convergence $z_n(s) \to \bar{z}(s)$ at each point of that partition and the inequality (1) for z_n imply that

$$\begin{aligned}\omega(\bar{z}, \Pi_N) &= \lim_{n\to\infty} \omega(z_n, \Pi_N) \\ &= \underline{\lim}_{n\to\infty}\omega(z_n, \Pi_N) \\ &\leq \underline{\lim}_{n\to\infty} \bigvee_a^b(z_n) \leq \bar{M}\end{aligned}$$

This provides reason enough to conclude that

$$\bigvee_a^b(\bar{z}) = \sup\left\{\omega(\bar{z}, \Pi_N) \colon \Pi_N \in \{\Pi_N\}\right\} \leq \underline{\lim}_{n\to\infty} \bigvee_a^b(z_n) \leq \bar{M}$$

as desired. □

Definition 1 *A sequence of functions $\{z_n(s)\} \subset V[a,b]$ is said to Π-converge to an element $z_0(s)$ if $z_n(s) \to z_0(s)$ as $n \to \infty$ for each $s \in [a,b]$ and $\|z_n\| \leq \bar{M} = const$ for all n.*

It is clear, from Lemma 1, that $z_0(s) \in V[a,b]$.

Lemma 2 *Let a sequence of functions $\{z_n(s)\} \subset V[a,b]$ Π-converge to a function $\bar{z}(s) \in V[a,b]$ as $n \to \infty$ and, in addition, $\|z_n\| \leq \|\bar{z}\|$ for all n. Then $\|z_n\| \to \|\bar{z}\|$ as $n \to \infty$.*

Proof. By Lemma 1 and the conditions of Lemma 2 we infer that

$$\|\bar{z}\| \leq \underline{\lim}_{n\to\infty}\|z_n\| \leq \overline{\lim}_{n\to\infty}\|z_n\| \leq \|\bar{z}\|$$

which implies the limit relation in question. □

Lemma 3 *Let the conditions of Lemma 2 hold for a sequence $\{z_n(s)\}$. Then*

$$\lim_{n\to\infty} \bigvee_{s_1}^{s_2}(z_n) = \bigvee_{s_1}^{s_2}(\bar{z})$$

without concern for how the numbers s_1, s_2 $(a \leq s_1 < s_2 \leq b)$ will be chosen.

Proof. Since the conditions of Lemma 2 (and consequently of Lemma 1) are satisfied, we take for granted that for arbitrary numbers s_1', s_2' $(a \leq s_1' < s_2' \leq b)$

$$\overline{\lim}_{n\to\infty} \bigvee_{s_1'}^{s_2'}(z_n) \geq \underline{\lim}_{n\to\infty} \bigvee_{s_1'}^{s_2'}(z_n) \geq \bigvee_{s_1'}^{s_2'}(\bar{z}) \tag{2}$$

Assume first, that for $s_1' = s_1$ and $s_2' = s_2$ the inequality in (2) is strict.

$$\overline{\lim}_{n\to\infty} \bigvee_{s_1}^{s_2}(z_n) > \bigvee_{s_1}^{s_2}(\bar{z}) \tag{3}$$

For the sake of definiteness, s_1 and s_2 will be taken to be interior points of the segment $[a, b]$. Then there exists a sequence of functions $\{z_{n_l}\} \subset \{z_n\}$ such that

$$\overline{\lim}_{n\to\infty} \bigvee_{s_1}^{s_2}(z_n) = \lim_{l\to\infty} \bigvee_{s_1}^{s_2}(z_{n_l}) > \bigvee_{s_1}^{s_2}(\bar{z}) \tag{4}$$

The norm boundedness condition $\|z_n\| \leq \|\bar{z}\|$ implies the boundedness of two numerical sequences

$$\left\{\bigvee_{a}^{s_1}(z_{n_l})\right\} \quad \left\{\bigvee_{s_2}^{b}(z_{n_l})\right\}$$

Hence one can choose a subsequence $\{z_{n_{l_r}}(s)\} \subset \{z_{n_l}(s)\}$ for which there exist the limits

$$\lim_{r\to\infty} \bigvee_{a}^{s_1}(z_{n_{l_r}}) \geq \bigvee_{a}^{s_1}(\bar{z}) \qquad \lim_{r\to\infty} \bigvee_{s_2}^{b}(z_{n_{l_r}}) \geq \bigvee_{s_2}^{b}(\bar{z}) \tag{5}$$

It is worth noting here that the lower estimates for these limits arose from (2). With the aid of relations (4)–(5) we arrive at

$$\begin{aligned} \lim_{r\to\infty} \bigvee_{a}^{b}(z_{n_{l_r}}) &= \lim_{r\to\infty} \left[\bigvee_{a}^{s_1}(z_{n_{l_r}}) + \bigvee_{s_1}^{s_2}(z_{n_{l_r}}) + \bigvee_{s_2}^{b}(z_{n_{l_r}})\right] \\ &> \bigvee_{a}^{s_1}(\bar{z}) + \bigvee_{s_1}^{s_2}(\bar{z}) + \bigvee_{s_2}^{b}(\bar{z}) = \bigvee_{a}^{b}(\bar{z}) \end{aligned}$$

Whence it follows that

$$\lim_{r\to\infty} \|z_{n_{l_r}}\| > \|\bar{z}\|$$

which disagrees with one of the conclusions of Lemma 2:

$$\lim_{n\to\infty} \|z_n\| = \lim_{r\to\infty} \|z_{n_{l_r}}\| = \|\bar{z}\|$$

Thus, if assumption (3) concerning the inequality strictness was imposed, we come to a contradiction. This supports the view that

$$\overline{\lim}_{n\to\infty} \bigvee_{s_1}^{s_2}(z_n) = \bigvee_{s_1}^{s_2}(\bar{z})$$

With this relation established, we finish the proof of the lemma by applying (2) to $s_1' = s_1$ and $s_2' = s_2$. □

Corollary 1 *Lemma 3 remains valid when a sequence $\{z_n\}$ Π-converges to $\bar{z}$ and*

$$\overline{\lim}_{n\to\infty}\|z_n\| \leq \|\bar{z}\|$$

Our further step is to state an important result which will be essential when establishing convergence of piece-uniform regularization algorithms.

Theorem 1 *Let $\{z_n(s)\} \subset V[a,b]$ be a sequence of functions such that*

(1) $\{z_n(s)\}$ Π-converges to $\bar{z}(s) \in V[a,b]$ as $n \to \infty$;

(2) $\|z_n\| \leq \|\bar{z}\|$ for each n.

Then $z_n(s)$ converges to $\bar{z}(s)$ uniformly on any segment $[s', s''] \subset [a,b]$ containing no discontinuity points of the limiting function $\bar{z}(s)$.

Proof. Suppose that Theorem 1 fails to be true. That is to say, on some segment $[s', s''] \subset [a,b]$ containing no discontinuity points of the function $\bar{z}(s)$ the convergence in question is not uniform, what means that for some $\varepsilon_0 > 0$ there exists a collection of points $s_n \in [s', s'']$ such that $|z_n(s_n) - \bar{z}(s_n)| > \varepsilon_0$ for $n \geq n_0$. In view of this, one can pick up from $\{s_n\}$ a convergent subsequence $\{s_{n_k}\}$ such that $s_{n_k} \to s^* \in [s', s'']$ as $k \to \infty$, whose use permits us to establish the chain of relations

$$\begin{aligned} \varepsilon_0 &< \left|z_{n_k}(s_{n_k}) - \bar{z}(s_{n_k})\right| \\ &\leq \left|z_{n_k}(s_{n_k}) - z_{n_k}(s^*)\right| \\ &\quad + \left|z_{n_k}(s^*) - \bar{z}(s^*)\right| \\ &\quad + \left|\bar{z}(s^*) - \bar{z}(s_{n_k})\right| \\ &\equiv J_{1k} + J_{2k} + J_{3k} \end{aligned} \tag{6}$$

It is necessary to examine the behavior of numerical sequences $\{J_{jk}\}_{k=1}^{\infty}$, $j = 1,2,3$, as $k \to \infty$. The continuity of the function $\bar{z}(s)$ on $[s', s'']$ and, in particular, at the point s^* provides the convergence $J_{3k} \to 0$. Under condition 1) of the theorem, $J_{2k} \to 0$. It remains to analyze more precisely only J_{1k} by appeal to an interior point s^* of the segment $[s', s'']$. When this is not the case, our arguments will be slightly different, but the main ideas behind the proof will be the same. If the function $\bar{z}(s) \in V[a,b]$ is continuous at the point $s^* \in (s', s'')$, then so is the function

$$\bar{q}(s) \equiv \bigvee_a^s(\bar{z})$$

by virtue of the well-known property of bounded variation functions (see, for example, Theorem 1.4.14). With this in mind, we take a small number $\delta_0 > 0$ so that $[s^* - \delta_0, s^* + \delta_0] \subset [s', s'']$ and the relations hold:

$$\bar{q}(s^* + \delta_0) - \bar{q}(s^* - \delta_0) = \bigvee_{s^*-\delta_0}^{s^*+\delta_0}(\bar{z}) < \frac{\varepsilon_0}{2} \tag{7}$$

Furthermore, we choose a number $N(\delta_0)$ in such a way to satisfy the inequality $|s^* - s_{n_k}| < \delta_0$ for all $k > N(\delta_0)$. In this way, we deduce with the aid of relation (1) that

$$\begin{aligned} J_{1k} &= \left| z_{n_k}(s_{n_k}) - z_{n_k}(s^*) \right| \\ &\le \left| \bigvee_{s_{n_k}}^{s^*} (z_{n_k}) \right| \le \bigvee_{s^*-\delta_0}^{s^*+\delta_0} (z_{n_k}) \qquad k > N(\delta_0) \end{aligned} \tag{8}$$

Observe that the conditions of Theorem 1 coincide with those of Lemmas 2 and 3. In view of this, applying Lemma 3 to the sequence $\{z_{n_k}\}$ yields

$$\lim_{k\to\infty} \bigvee_{s^*-\delta_0}^{s^*+\delta_0} (z_{n_k}) = \bigvee_{s^*-\delta_0}^{s^*+\delta_0} (\bar{z}) \tag{9}$$

Hence, passing in (8) to the limit and keeping in mind (7) and (9), we obtain

$$\lim_{k\to\infty} J_{1k} < \frac{\varepsilon_0}{2}$$

The collection of the estimates just derived from formula (6) for convergences of J_{jk}, $j = 1, 2, 3$, constitutes the following inconsistent inequalities

$$\varepsilon_0 \le \lim_{k\to\infty} \left(J_{1k} + J_{2k} + J_{3k} \right) < \frac{\varepsilon_0}{2}$$

This completes the proof of the theorem. □

Corollary 2 *Let condition* 1) *of Theorem* 1 *hold and let condition* 2) *of Theorem* 1 *be replaced by the limit relation*

$$\overline{\lim}_{n\to\infty} \|z_n\| \le \|\bar{z}\|$$

Then the assertion of Theorem 1 *is still valid.*

Proof. Indeed, from the inequality proved in Lemma 1 and the condition of Corollary 2 it seems clear that the convergence $\|z_n\| \to \|\bar{z}\|$ occurs as $n \to \infty$. Hence Lemma 3 being actually used in the proof of Theorem 1 is valid due to Corollary 1. □

We conclude our exposition with a discussion of one property of bounded variation functions.

Definition 2 *A discontinuity of the first kind at point* $s_0 \in (a, b)$ *of a function* $z(s) \in V[a, b]$ *is called usual if*

$$\begin{aligned} \underline{\lim}_{s\to s_0} z(s) &= \min\left[z(s_0 - 0), z(s_0 + 0)\right] \le z(s_0) \\ &\le \overline{\lim}_{s\to s_0} z(s) = \max\left[z(s_0 - 0), z(s_0 + 0)\right] \end{aligned}$$

When a and b happen to be discontinuity points of the function $z(s)$, they are supposed to be usual by definition.

It is obvious that a discontinuity point $s_0 \in (a, b)$ is usual if and only if

$$\left|z(s_0 - 0) - z(s_0)\right| + \left|z(s_0 + 0) - z(s_0)\right| = \left|z(s_0 - 0) - z(s_0 + 0)\right|$$

Lemma 4 *Let functions $z_1(s)$ and $z_2(s)$ from the space $V[a,b]$ coincide everywhere except maybe their common discontinuity points from (a,b) being usual ones of the function $z_1(s)$. Then $\|z_1\| = \|z_2\|$ if all of the discontinuity points of the function $z_2(s)$ are usual and $\|z_1\| < \|z_2\|$ otherwise.*

Proof. It suffices to consider the case when $z_1(s)$ and $z_2(s)$ differ at one of the discontinuity points, say s_0 $(a < s_0 < b)$. For any function z from the space $V[a,b]$ we might have

$$\begin{aligned}\bigvee_a^b(z) &= \lim_{s\to s_0-0}\bigvee_a^s(z) + \lim_{s\to s_0+0}\bigvee_s^b(z)\\ &\quad+\big|z(s_0-0)-z(s_0)\big|\\ &\quad+\big|z(s_0+0)-z(s_0)\big|\end{aligned}$$

The resulting equality for $z = z_1$ and the relation

$$\big|z_1(s_0-0)-z_1(s_0)\big| + \big|z_1(s_0+0)-z_1(s_0)\big| = \big|z_1(s_0-0)-z_1(s_0+0)\big|$$

which is valid at every usual discontinuity point, imply that

$$\begin{aligned}\bigvee_a^b(z_1) &= \lim_{s\to s_0-0}\bigvee_a^s(z_1) + \lim_{s\to s_0+0}\bigvee_s^b(z_1)\\ &\quad+\big|z_1(s_0-0)-z_1(s_0+0)\big|\end{aligned}$$

The function $z_2(s)$ coincides with $z_1(s)$ everywhere except the point s_0. In just the same way as before, one can derive the useful expressions

$$\begin{aligned}\bigvee_a^b(z_2) &= \lim_{s\to s_0-0}\bigvee_a^s(z_2) + \lim_{s\to s_0+0}\bigvee_s^b(z_2)\\ &\quad+\big|z_2(s_0-0)-z_2(s_0)\big|\\ &\quad+\big|z_2(s_0+0)-z_2(s_0)\big|\\ &= \lim_{s\to s_0-0}\bigvee_a^s(z_1) + \lim_{s\to s_0+0}\bigvee_s^b(z_1)\\ &\quad+\big|z_2(s_0-0)-z_2(s_0)\big|\\ &\quad+\big|z_2(s_0+0)-z_2(s_0)\big|\end{aligned}$$

We know that the inequality

$$\begin{aligned}&\big|z_2(s_0-0)-z_2(s_0)\big| + \big|z_2(s_0+0)-z_2(s_0)\big|\\ &\geq \big|z_2(s_0-0)-z_2(s_0+0)\big|\\ &= \big|z_1(s_0-0)-z_1(s_0+0)\big|\end{aligned}$$

would be strict if s_0 is none of the usual discontinuity points of $z_2(s)$. Thus, the equality sign is attained when s_0 is the usual discontinuity point of $z_2(s)$. Finally, by comparing the variations of the functions z_1 and z_2, we arrive at the statement of Lemma 4. □

Remark The assertions of Theorem 1 and Lemmas 1–3 type have been considered in a more common setting by Leonov (1982 a) (compare with Goncharskiĭ and Stepanov (1979), Tikhonov *et al.* (1983)).

5.2 The statement of the problem and algorithms for its solution

Let A be an operator from the space $V[a,b]$ into a normed space U and let F be a functional on $V[a,b]$.

Definition 1 *An operator A (a functional F) is said to be Π-continuous on an element $z_0 \in V[a,b]$ if $Az_n \xrightarrow{U} Az_0$ ($F(z_n) \to F(z_0)$) for any sequence $\{z_n(s)\}$, which Π-converges to $z_0(s)$ as $n \to \infty$. If A (or F) is Π-continuous for every $z_0 \in V[a,b]$, we say that it is Π-continuous on $V[a,b]$.*

We spoke above about the functional space $Z \equiv V[a,b]$. For the purposes of the present section, we endow Z with the topology τ of Π-convergence. Let A be a Π-continuous operator on Z and $\Omega_1(z)$ be a nonnegative Π-continuous functional on Z. Of special interest will be the functional $\Omega(z) \equiv \|z\| + p\Omega_1(z)$, where $p \geq 0$ is a constant.

Lemma 1 *The functional $J(z) \equiv \|Az - u\|_U$ is τ-continuous on Z, while the functional $\Omega(z)$ is lower τ-semicontinuous on Z.*

Proof. Observe that if the operator A is Π-continuous, then so is the functional J. Consider an arbitrary sequence $\{z_n(s)\} \subset V[a,b]$, which Π-converges to $z_0(s) \in V[a,b]$. From Π-continuity of the functional Ω_1 we derive by Lemma 1.1 the relations

$$\begin{aligned}\underline{\lim}_{n\to\infty}\Omega(z_n) &= \underline{\lim}_{n\to\infty}\{\|z_n\| + p\Omega_1(z_n)\}\\ &= \underline{\lim}_{n\to\infty}\|z_n\| + p\lim_{n\to\infty}\Omega_1(z_n)\\ &\geq \|z_0\| + p\Omega_1(z_0) = \Omega(z_0)\end{aligned}$$

which mean the lower semicontinuity of the functional Ω with respect to Π-convergence on Z. □

Lemma 2 *Any nonempty set of the form $\Omega_C \equiv \{z \in V[a,b] \colon \Omega(z) \leq C\}$ is τ-sequentially compact.*

Proof. Indeed, the inequality $\Omega(z_n) \leq C$, which is valid for any sequence $\{z_n\} \subset \Omega_C$, provides the uniform boundedness of the norm: $\|z_n\| \leq C$. By Helly's theorem on compactness (see Theorem 1.4.15) the sequence $\{z_n\}$ is compact with respect to pointwise convergence and, therefore, τ-sequentially compact too. □

Let us turn to an operator equation of the form

$$Az = u \qquad z \in V[a,b] \quad u \in U \tag{1}$$

and assume that the set of its pseudosolutions Z^* corresponding to the right-hand side $\bar{u} \in U$ is nonempty.

Observe that the set $\bar{Z}$ of Ω-optimal pseudosolutions to equation (1) on $V[a,b]$ appears to be nonempty. This fact follows immediately from Theorem 2.4.1 and Lemmas 1–2.

Now, we will assume that the **generalized uniqueness property** of Ω-optimal pseudosolutions is true, saying that all the Ω-optimal pseudosolutions to equation (1) pointwise coincide on $[a,b]$ except maybe their common discontinuity points. They are supposed to be usual, if any. According to Lemma 1.5, this property causes the norm coincidence of all Ω-optimal pseudosolutions and, in view of the equalities

$$\Omega(z) = \|z\| + p\Omega_1(z) \qquad \Omega(z) = \bar{\Omega} \qquad \forall z \in \bar{Z}$$

to the value constancy of the functional Ω_1 on the set $\bar{Z}$. To facilitate subsequents arguments, we take one of the Ω-optimal solutions and denote it by $\bar{z}(s)$.

In mastering difficulties involved we try to involve the approximations (A_h, u_σ) following the assumptions of Section 3.1 in Chapter 3 with no exact data $(A, \bar{u})$ of problem (1). This means that the operator A_h defined on $V[a,b]$ is Π-continuous and satisfies the condition of approximation

$$\|A_h z - Az\|_U \leq \psi(h, \|z\|) \qquad \forall z \in V[a,b]$$

where the number $h \geq 0$ and the function ψ are known, and ψ possesses the standard properties listed in Section 2.2 in Chapter 2. We have at our disposal a known number $\sigma \geq 0$ and an element $u_\sigma \in U$ with $\|\bar{u} - u_\sigma\|_U \leq \sigma$.

The problem on piece-uniform regularization of equation (1) is to construct from the available data $(A_h, u_\sigma, h, \sigma, \psi)$ approximations $z_\delta \in V[a,b]$ $(\delta \equiv (h,\sigma))$ to Ω-optimal pseudosolutions from the set $\bar{Z}$, so that for any sequence $\{\delta_n\}$ tending to 0 as $n \to \infty$ the appropriate sequence $\{z_n(s)\}$, $z_n \equiv z_{\delta_n}$, converges to $\bar{z}(s)$ uniformly on any segment $[s', s''] \subset [a,b]$ containing no discontinuity points of the function $\bar{z}(s)$.

Since the assumptions of Section 3.1 in Chapter 3 are satisfied for the problem statement described above, we can employ the **g.p.d.**, **g.p.q.** and **g.p.f.** algorithms of Section 3.2 in Chapter 3 for constructing Ω-optimal pseudosolutions to equation (1). Here, they are based on minimizing a smoothing functional of the form

$$\begin{aligned} M^\alpha[z] &= \alpha g[|z(a)| + \bigvee_a^b(z) + p\Omega_1(z)] \\ &\quad + f(\|A_h z - u_\sigma\|) \quad \alpha > 0 \quad z \in Z \end{aligned}$$

where the function $g(x)$ defined for $x \geq 0$ is continuous and monotonically increasing with $g(+\infty) = +\infty$ and $f(x) \in \mathcal{F}^m[0, +\infty)$. Let $z^{\alpha_\delta} \in V[a,b]$ be an approximation to $\bar{Z}$ obtained by means of the algorithms mentioned. Then, in the framework of the general theory outlined in Chapters 2–3, for any sequence $\{\delta_n\}$ tending to 0 as $n \to \infty$ the appropriate sequence of approximate solutions $\{z_n\}$, $z_n = z^{\alpha_{\delta_n}}$, П-converges to $\bar{Z}$ as $n \to \infty$ and $\Omega(z_n) \to \bar{\Omega}$. Moreover, it turns out that the approximations z^{α_δ} are just a solution of the piece-uniform regularization problem. In this regard, the following theorem is valid.

Theorem 1 *The sequence $\{z_n(s)\}$ converges to $\bar{z}(s)$ as $n \to \infty$ uniformly on any segment $[s', s''] \subset [a, b]$ containing no discontinuity points of the function $\bar{z}(s)$.*

Proof. Since the sequence $\{z_n\}$ П-converges to $\bar{Z}$ and all the elements of the set $\bar{Z}$ have the same norm, one can deduce by Lemma 1.1 from the П-continuity of the functional Ω_1 and the convergence $\Omega(z_n) \to \bar{\Omega}$ as $n \to \infty$ that

$$\begin{aligned} \|\bar{z}\| &\le \underline{\lim}_{n\to\infty} \|z_n\| \le \overline{\lim}_{n\to\infty} \|z_n\| \\ &= \overline{\lim}_{n\to\infty} \left[\Omega(z_n) - p\Omega_1(z_n)\right] \\ &= \lim_{n\to\infty} \Omega(z_n) - p \lim_{n\to\infty} \Omega_1(z_n) \\ &= \bar{\Omega} - p\Omega_1(\bar{z}) \\ &= \Omega(\bar{z}) - p\Omega_1(\bar{z}) = \|\bar{z}\| \end{aligned}$$

exploiting the fact that $\Omega(z)$ and $\Omega_1(z)$ both preserve their values on $\bar{Z}$. As a final result, we establish that $\|z_n\| \to \|\bar{z}\|$ as $n \to \infty$.

Assume to the contrary that the sequence $\{z_n\}$ does not converge uniformly to $\bar{z}(s)$ on some segment $[s', s''] \subset [a, b]$ containing no discontinuity points of the function $\bar{z}(s)$. This means that there exists a subsequence $\{z_{n_k}\} \subset \{z_n\}$ such that

$$\Delta_k \equiv \sup\left\{|z_{n_k}(s) - \bar{z}(s)| \colon s \in [s', s'']\right\} \ge \varepsilon_0 = \text{const} > 0 \qquad (2)$$

for any subscript k. Since the norms $\|z_{n_k}\|$ are uniformly bounded, there exists, due to Helly's theorem on compactness, a subsequence $\{z_r\} \subset \{z_{n_k}\}$, $r \equiv n_{k_l}, l = 1, 2, \ldots$, which converges to some function $z^*(s) \in V[a, b]$ at each point $s \in [a, b]$ and, consequently, П-converges to $z^*(s)$. Because the set $\bar{Z}$ is П-closed (see Corollary 2.4.1), the function $z^*(s)$ gives an Ω-optimal pseudosolution. In accordance with what has been said the convergences

$$z_r \xrightarrow{\Pi} z^* \text{ and } \|z_r\| \to \|\bar{z}\| = \|z^*\|$$

just established as $r \to \infty$ provide that the sequence $\{z_r\}$ satisfies the conditions of Theorem 2.1 and, therefore, converges to $z^*(s)$ uniformly over any segment containing no discontinuity points of the function $z^*(s)$. Being an element of $\bar{Z}$, $z^*(s)$ coincides with $\bar{z}(s)$ everywhere except their common discontinuity points. In light of the uniform convergence of $\{z_r\}$

on the segment $[s', s'']$, we would have

$$\begin{aligned} \Delta_r &= \sup\{|z_r(s) - \bar{z}(s)| : s \in [s', s'']\} \\ &= \sup\{|z_r(s) - z^*(s)| : s \in [s', s'']\} \to 0 \end{aligned}$$

as $r \to \infty$. But this contradicts Assumption (2) and thereby completes the proof of the theorem. □

Next, the subsidiary *a priori* information on an Ω-optimal pseudosolution $\bar{z}(s)$ to equation (1) is concerned with extra properties of the solution:

(A) The solution is piece-monotone and perhaps has a finite number of usual discontinuity points of the first kind with unknown positions.

(B) The solution has a finite number of usual discontinuity points of the first kind with unknown positions and possesses a finite derivative of class $L_q[a, b]$ $(q \geq 1)$ everywhere on $[a, b]$ except discontinuity points.

(C) The solution is differentiable everywhere on $[a, b]$ and the derivative is q-summable with $q \geq 1$.

(D) The solution is differentiable and admits a finite number of local minima and maxima with unknown positions.

Under this agreement, we know from Kolmogorov and Fomin (1968) and Natanson (1974) that the inclusion $\bar{z}(s) \in V[a, b]$ should occur and, according to Theorem 1, the **g.p.d.**, **g.p.q.** and **g.p.f.** algorithms provide piece-uniform convergence of approximations in the cases (A)–(B) and uniform convergence on the segment $[a, b]$ in the cases (C)–(D).

We focus our attention on finite-dimensional approximation of the problem under consideration. As before, it is convenient to introduce on the segment $[a, b]$ the family of grids $\Pi_N = \{a = s_1 < \cdots < s_N = b\}$, $N \geq 2$, and a finite-dimensional space $Z_N = \mathbf{R}^N$ the elements of which $\hat{z}_N = (z_1, \ldots, z_N)$ give the values of a grid function at the nodes of the grid Π_N. Arguing as in Sections 4.1 and 4.4 in Chapter 4, we refer to an operator P_N acting from $V[a, b]$ into Z_N and being the projector onto the grid: $P_N z = (z(s_1), \ldots, z(s_N))$ for every $z(s) \in V[a, b]$.

Furthermore, for each grid Π_N, an operator $\bar{P}_N \colon Z_N \to V[a, b]$ is treated as the operator of filling values of a grid function between the grid nodes. Let that operator possess the following properties:

(1) $\bar{P}_N \colon Z_N \to V[a, b]$ is a linear operator.

(2) $z_N(s_i) \equiv (\bar{P}_N \hat{z}_N)(s_i) = z_i$ for each $s_i \in \Pi_N$ $(i = 1, \ldots, N)$ and any element $\hat{z}_N = (z_1, \ldots, z_N) \in \mathbf{R}^N$.

(3) The function $z_N(s)$ is monotone on every segment $[s_i, s_{i+1}]$ $(i = 1, \ldots, N - 1)$.

Any such operator $\bar{P}_N$ is called a **monotone filling operator**.

It is not difficult to see that properties (2)–(3) imply

$$\begin{aligned}\|\bar{P}_N\hat{z}_N\| \equiv \|z_N(s)\| &= |(\bar{P}_N\hat{z}_N)(s_1)| \\ &\quad + \sum_{k=1}^{N-1}\bigvee_{s_k}^{s_{k+1}}(\bar{P}_N\hat{z}_N) \\ &= |z_1| + \sum_{k=1}^{N-1}|z_{k+1} - z_k| \end{aligned} \tag{3}$$

Observe that the expression in the right-hand side of the preceding formula is equivalent to the norm of the space $\mathbf{R}^N$. This provides support for the view that the linear operator $\bar{P}_N$ is bounded for each fixed N and, therefore, $\bar{P}_N$ is a continuous operator on $\mathbf{R}^N$ into $V[a,b]$.

It is evident that Assumptions 1–3 of Section 4.1 in Chapter 4 are satisfied for the operators P_N and $\bar{P}_N$ in accordance with their definition and properties.

The linear interpolation operator can serve as one possible example of the operator $\bar{P}_N$. Another example is the operator $\bar{P}_N^*$ carrying every vector $\hat{z}_N \in \mathbf{R}^N$ into a step function

$$z_N^*(s) = (\bar{P}_N^*\hat{z}_N)(s) = \begin{cases} z_i & s_i \le s < s_{i+1}, i = 1,\dots, N-1 \\ z_N & s = s_N \end{cases} \tag{4}$$

It is worth emphasizing here that the function $z_N(s) = (\bar{P}_N\hat{z}_N)(s)$ may have discontinuity points of the first kind.

To satisfy Assumption 4 of Section 4.1 in Chapter 4 imposed on the operators P_N and $\bar{P}_N$ we assume that the functional Ω_1 possesses, in addition to the properties listed above, a characteristic feature:

$$\overline{\lim}_{n\to\infty}\Omega_1(\bar{P}_N P_N\bar{z}) \le \Omega_1(\bar{z}) \qquad \forall \bar{z} \in \bar{Z}$$

Then, in view of (1.1) and (3), we arrive at

$$\begin{aligned}\overline{\lim}_{N\to\infty}\Omega(\bar{P}_N P_N\bar{z}) &= \overline{\lim}_{N\to\infty}\Big\{|\bar{z}(s_1)| \\ &\quad + \sum_{k=1}^{N-1}\big|\bar{z}(s_{k+1}) - \bar{z}(s_k)\big| + p\Omega_1(\bar{P}_N P_N\bar{z})\Big\} \\ &\le \overline{\lim}_{N\to\infty}\{|\bar{z}(s_1)| + \bigvee_a^b(\bar{z}) + p\Omega_1(\bar{P}_N P_N\bar{z})\} \\ &\le \|\bar{z}\| + p\Omega_1(\bar{z}) = \Omega(\bar{z})\end{aligned}$$

thereby justifying that Assumption 4 of Section 4.1 in Chapter 4 holds true.

Finally, still using the framework of Section 4.4 in Chapter 4, consider a continuous operator $\bar{Q}_M$ acting from a finite-dimensional space U_M of the dimension M into U. For a moment, instead of the operator A_h we deal with a finite-dimensional continuous operator $\hat{A}_H: \mathbf{R}^N \to U_M$, $H \equiv (h, 1/M, 1/N)$, such that the condition of approximation

$$\|A_h z - \bar{Q}_M \hat{A}_H P_N z\|_U \le \widehat{\psi}(H, \|z\|) \qquad \forall z \in V[a,b] \tag{5}$$

holds true. The measure of finite-dimensional approximation $\widehat{\psi}$ satisfying Assumption 3 of Section 2.2 in Chapter 2 is supposed to be known. As will be shown in Section 3, it can be found for plenty of problems in many ways.

Instead of u_σ, let us consider its finite-dimensional approximation $\hat{u}_{\sigma M} \in U_M$ satisfying the condition of approximation

$$\|u_\sigma - \bar{Q}_M \hat{u}_{\sigma M}\|_U \le \xi(1/M, \sigma)$$

where the function ξ is known and obeys the properties of Section 4.4 in Chapter 4.

This provides reason enough to conclude that all of the conditions of the statement of the finite-dimensional approximation problem related to equation (1) in the framework of Sections 4.1 and 4.4 in Chapter 4 are satisfied and to employ the finite-dimensional **g.p.d.**, **g.p.q.** and **g.p.f.** algorithms for the approximate determination of Ω-optimal pseudosolutions to equation (1) on the basis of available data

$$\hat{A}_H, \hat{u}_{\sigma M}, H = (h, 1/M, 1/N), \sigma, \psi(h, \|z\|), \widehat{\psi}(H, \|z\|)$$
$$\xi(1/M, \sigma), M, N, U_M, P_N, \bar{P}_N, \bar{Q}_M$$

Common practice involves minimizing the smoothing functional (4.1.5) taking for now the form

$$\begin{aligned} \widehat{M^\alpha}[\hat{z}_N] &= \alpha g\Big\{|z_1| + \sum_{k=1}^{N-1} |z_{k+1} - z_k| + p\Omega_1(\bar{P}_N \hat{z}_N)\Big\} \\ &\quad + f(\|\bar{Q}_M \hat{A}_H \hat{z}_N - \bar{Q}_M \hat{u}_{\sigma M}\|_U) \\ &\quad \alpha > 0 \qquad \hat{z}_N \in \mathbf{R}^N \end{aligned} \tag{6}$$

In dealing with approximations $z_\eta(s) \equiv \bar{P}_N \hat{z}_N^{\alpha_\eta}$, $(\eta \equiv (H, \sigma))$ obtained by means of the algorithms mentioned above the propositions of Sections 4.2–4.4 in Chapter 4 justify that for any sequence $\{\eta_n\}$ such that $\eta_n \to 0$ as $n \to \infty$, the appropriate sequence of approximations $\{z_n(s)\}$, $z_n \equiv z_{\eta_n}$, does follow the convergence properties

$$z_n(s) \xrightarrow{\Pi} \bar{Z} \text{ and } \Omega(z_n) \to \bar{\Omega} \text{ as } n \to \infty$$

By exactly the same reasoning as in Theorem 1, one can draw the conclusion that the sequence $\{z_n(s)\}$ possesses the property of piece-uniform regularization, that is, $z_n(s)$ converges to $\bar{z}(s)$ uniformly on any segment $[s', s''] \subset [a,b]$ containing no discontinuity points of the function $\bar{z}(s)$.

Remark The problem on piece-uniform regularization was first considered in other settings by Arsenin (1965) and Goncharskiĭ and Yagola (1969) and later by Leonov (1980, 1982 a). The functions of bounded variation were widely used in the study of ill-posed problems from other points of view in the works of Ageev (1980), Dmitriev and Poleshchuk (1972), Zagonov (1987) without solving the problem of piece-uniform regularization. One possible variant of the algorithm of the generalized discrepancy method with the intervention of bounded variation functions was investigated by Goncharskiĭ and Stepanov (1979, 1980) and Tikhonov *et al.* (1983) who showed that such algorithms provide uniform regularization of continuous solutions.

5.3 Examples

This section discusses frequently encountered operator equations whose piece-uniform regularization can be realized in the framework of Section 2.

Example 1 Let A be an integral operator of the form

$$A[x, z(s)] = \int_a^b K(x, s)dz(s) \qquad x \in [c, d] \tag{1}$$

where $K(x, s) \in C^1(\overline{\Pi})$, $\overline{\Pi} \equiv [c, d] \times [a, b]$ and $z(s) \in V[a, b]$. As known, any such operator is defined on the space $V[a, b]$. We claim that it is Π-continuous on $V[a, b]$ into $L_2[c, d]$ too and proceed as usual. This amounts to considering an arbitrary sequence $\{z_n(s)\} \subset V[a, b]$, which Π-converges to $z_0(s) \in V[a, b]$, and introducing

$$u_n(x) \equiv A[x, z_n(s)] - A[x, z_0(s)] = \int_a^b K(x, s)d[z_n(s) - z_0(s)]$$

As a matter of fact, the sequence $\{u_n(x)\}$ possesses the following properties:

(a) It is uniformly bounded on $[c, d]$ by Riesz's theorem (see Theorem 1.4.16):

$$\begin{aligned} |u_n(x)| &= \left| \int_a^b K(x, s)d[z_n(s) - z_0(s)] \right| \\ &\le \|K\|_C \cdot \|z_n - z_0\| \le \|K\|_C\big(\|z_n\| + \|z_0\|\big) \le \text{const} \\ &\quad u_n(x) \in C[c, d] \end{aligned}$$

(b) It pointwise converges to 0 on the segment $[c, d]$ as $n \to \infty$ due to

Helly's theorem on the passage to the limit (see Theorem 1.4.17):

$$|u_n(x)| = \left| \int_a^b K(x,s)dz_n(s) - \int_a^b K(x,s)dz_0(s) \right| \to 0$$

$$n \to \infty \qquad \forall x \in [c,d]$$

By the Lebesgue theorem (see corollary to Theorem 1.4.8) properties (a)–(b) imply that

$$\|u_n\|^2_{L_2} = \int_c^d |u_n(x)|^2 dx \to \int_c^d 0\, dx = 0 \quad n \to \infty$$

which justifies Π-continuity of the operator A on $V[a,b]$ into $L_2[c,d]$.

Of special interest is the following equation of the form (2.1):

$$A[x, z(s)] = u(x) \qquad u(x) \in L_2[c,d] \equiv U \tag{2}$$

It is easy to show that if it has pseudosolutions from the space $V[a,b]$, then one might expect the existence of infinitely many pseudosolutions. In fact, the values of integral (1) do not depend on the values of the function $z(s)$ at its discontinuity points (see Section 1.4.8 in Chapter 1) and, in particular, at removable discontinuity points. Therefore, 'having spoilt' a pseudosolution to equation (2) by removable discontinuity points, it is possible to obtain as many pseudosolutions as desired. They will be different as the elements of the space $V[a,b]$.

However, by Lemma 1.4, a normal pseudosolution to equation (2) has no removable discontinuity points whereas its discontinuity points of the first kind are usual. Everything just said applies equally well to any Ω-optimal pseudosolution if the functional Ω_1 takes the form (4) (see below) or certain other forms.

In a number of possible cases, Ω-optimal solutions to equation (2) possess the generalized uniqueness property. In particular, we will assume that this property is valid for the right-hand side $\bar{u}(x) \in C^1[c,d]$.

We confine ourselves to the case where operator (1) and $\bar{u}(x)$ are representable in the explicit forms ($h = 0, \sigma = 0$) and their finite-dimensional approximations are available. It is straightforward to verify that the corresponding conditions of Section 2 are satisfied.

Let

$$\Pi_N = \{s_k \colon s_k = a + \tau_N(k-1), k = 1, \ldots, N\}$$

be an equidistant grid on $[a,b]$ with spacing $\tau_N = (b-a)/(N-1)$. Likewise, we introduce on $[c,d]$ a similar grid

$$\Pi_M = \{x_i \colon x_i = c + \tau_M(i-1), i = 1, \ldots, M\}$$

with spacing $\tau_M = (d-c)/(M-1)$. We assume that

$$Z_N = \mathbf{R}^N \text{ and } U_M = \mathbf{R}^M$$

are finite-dimensional spaces of all grid functions defined on the grids Π_N and Π_M, respectively. As the operators $\bar{P}_N$ and $\bar{Q}_M$ we agree to consider step filling operators of the form (2.4) on the corresponding grids and refer to a finite-dimensional operator $\hat{A}_H$ $(H = (1/M, 1/N))$ approximating operator (1) in accordance with the rule

$$\begin{aligned}
&\hat{A}_H \hat{z}_N \equiv \sum_{j=1}^{N-1} K(x_i, s_j)\Delta z_j \\
&x_i \in \Pi_M \quad s_j \in \Pi_N \quad \Delta z_j = z_{j+1} - z_j \\
&i = 1, \dots, M \quad j = 1, \dots, N-1
\end{aligned}$$

Our further step is to derive a condition of finite-dimensional approximation.

The definition of the operators $\bar{Q}_M$, P_N and the estimate

$$|K(x,s) - K(x_i, s_j)| \leq \|K'_x\|_C \cdot \tau_M + \|K'_s\|_C \cdot \tau_N$$

which is valid for $x \in [x_i, x_{i+1}]$ and $s \in [s_j, s_{j+1}]$ $(i = 1, \dots, M-1, j = 1, \dots, N-1)$, imply that

$$\begin{aligned}
&\left|A[x, z(s)] - \bar{Q}_M \widehat{A}_H P_N z\right| \\
&= \left|\sum_{j=1}^{N-1} \left[\int_{s_j}^{s_{j+1}} K(x,s) dz - K(x_i, s_j) \int_{s_j}^{s_{j+1}} dz\right]\right| \\
&= \left|\sum_{j=1}^{N-1} \int_{s_j}^{s_{j+1}} \left[K(x,s) - K(x_i, s_j)\right] dz\right| \\
&\leq \sum_{j=1}^{N-1} \left\|K(x,s) - K(x_i, s_j)\right\|_{C(\Pi_{ij})} \cdot \bigvee_{s_j}^{s_{j+1}} (z) \\
&\leq \left(\|K'_x\|_C \tau_M + \|K'_s\|_C \tau_N\right) \sum_{j=1}^{N-1} \bigvee_{s_j}^{s_{j+1}} (z) \\
&\leq \left(\|K'_x\|_C \tau_M + \|K'_s\|_C \tau_N\right) \|z\| \qquad (3)
\end{aligned}$$

keeping $x \in [x_i, x_{i+1}]$, $\Pi_{ij} \equiv [x_i, x_{i+1}] \times [s_j, s_{j+1}]$ and using the estimate from Riesz's theorem (see Section 1.4.8 in Chapter 1)

$$\left|\int_a^b f(x) dz(x)\right| \leq \|f\|_{C[a,b]} \cdot \|z\|_{V[a,b]} \quad f \in C[a,b] \quad z \in V[a,b]$$

It follows from the foregoing that

$$\left\|A[x, z(s)] - \bar{Q}_M \widehat{A}_H P_N z\right\|_{L_2[c,d]} \leq \widehat{\psi}(H, \|z\|)$$
$$= \sqrt{d-c}\left(\|K'_x\|_C \tau_M + \|K'_s\|_C \tau_N\right)\|z\|$$

which represents an alternative form of the approximation condition (2.5). The question of approximation of the right-hand side has been investigated in considerable detail in Section 4.5 in Chapter 4.

We now dwell on the problem of making a choice of the functional $\Omega_1(z)$ satisfying the requirements of Section 2. For example, having defined the functional

$$\Omega_1(z) = z^2(b) + \int_a^b z^2(s)ds \tag{4}$$

on $V[a, b]$ as on the subset of the space $L_2[a, b]$ we will be sure, due to the Lebesgue theorem, that it is Π-continuous. The validity of the condition

$$\lim_{N \to \infty} \Omega_1(\bar{P}_N P_N \bar{z}) = \Omega_1(\bar{z}) \qquad \forall \bar{z} \in \bar{Z}$$

is easily verified. Observe that its variant has been already used in Section 2.

Indeed, for the step filling operator $\bar{P}_N$ we might have

$$\Omega_1(\bar{P}_N P_N z) = z^2(b) + \sum_{k=1}^{N-1} z^2(s_k)\tau_N \qquad \forall z \in V[a, b]$$

where the integral sum for the function $z^2(s) \in V[a, b]$ in the right-hand side converges to the integral involved in formula (4) as $N \to \infty$ because $z^2(s)$ is summable.

From such reasoning, it seems clear that the **g.p.d.**, **g.p.q.** and **g.p.f.** algorithms are much applicable for establishing piece-uniform regularization in the framework of Section 2 of various equations having the form (2).

Example 2 Set

$$A[x, z(s)] = \int_a^b K(x, s, z(s))ds \quad x \in [c, d] \quad z(s) \in V[a, b] \tag{5}$$

where the function $K(x, s, z) \in C(\Pi_0)$ ($\Pi_0 = [c, d] \times [a, b] \times \mathbf{R}$) satisfies the conditions of Example 4.5.1. By the Lebesgue theorem, it follows from the inequality

$$\left|K(x, s, z_1(s)) - K(x, s, z_2(s))\right| \leq \xi_3(z_0)|z_1(s) - z_2(s)|$$
$$z_0 \equiv \max\left\{\|z_1\|, \|z_2\|\right\}$$

arising from Example 4.5.1 that operator (5) is Π-continuous on $V[a, b]$ and acts into $L_2[c, d]$. Finite-dimensional approximation of operator (5) is carried out in a similar manner to that used in Example 1 by merely inserting

$$\widehat{A}_H \hat{z} = \sum_{j=1}^{N} K(x_i, s_j, z_j)\tau_N \quad x_i \in \Pi_M \quad s_j \in \Pi_N$$

$$i = 1, \dots, M \quad j = 1, \dots, N \quad \hat{z} = (z_1, \dots, z_N) \in Z_N$$

Following the instructions of Example 4.5.1 and adopting the techniques of Example 1 we derive the condition of approximation in the form

$$\begin{aligned} &\left\| A[x, z(s)] - \bar{Q}_M \widehat{A}_H P_N z \right\|_{L_2[c,d]} \le \widehat{\psi}(H, \|z\|) \\ &= \sqrt{d-c}\big\{ [\xi_0(\|z\|) + \xi_2(\|z\|)(b-a)/2 \\ &\quad + \xi_3(\|z\|)\|z\|]\tau_N + \xi_1(\|z\|)(b-a)\tau_M \big\} \end{aligned}$$

Of course, there are other examples of problems that are consistent with the theory of Section 2 and other ways of constructing finite-dimensional approximations of the problems from Examples 1–2.

5.4 Realization of piece-uniform regularization algorithms

The piece-uniform regularization algorithms whose finite-dimensional variants were presented in Section 2 are based on solving the problem of minimizing a smoothing functional in which it is necessary to obtain an element $\hat{z}^\alpha \in \mathbf{R}^N$ such that

$$\widehat{M}^\alpha[\hat{z}^\alpha] = \inf\left\{ \widehat{M}^\alpha[\hat{z}] : \hat{z} \in \mathbf{R}^N \right\} \qquad \hat{z} = (z_1, \dots, z_N) \tag{1}$$

if we have at our disposal

$$\begin{aligned} &\widehat{M}^\alpha[\hat{z}] = \alpha\left\{ |z_1| + \textstyle\sum_{k=1}^{N-1} |z_{k+1} - z_k| \right\} + \Phi_\alpha(\hat{z}) \\ &\Phi_\alpha(\hat{z}) \equiv \alpha p \Omega_1(\bar{P}_N \hat{z}) + f(\|\bar{Q}_M \widehat{A}_H \hat{z} - \bar{Q}_M \hat{u}_{\sigma M}\|_U) \end{aligned} \tag{2}$$

(see (2.6)).

Some difficulties do arise in the minimization path of the nonsmooth finite-dimensional functional (2) so, to assist the readers, we offer below one useful procedure by means of which the unconditional minimization problem (1) with the nonsmooth functional (2) can be reduced to another problem with a special differentiable functional under simple restrictions.

Here, we use a set of all vectors $\widehat{Z}\uparrow$ of the space $\mathbf{R}^N$ the components of which are arranged in the nondecreasing order: $z_k \le z_{k+1}$, $k = 1, \dots, N-1$. The introduction of new vectors $\hat{v}, \hat{w} \in \mathbf{R}^N$ by the relations

$$\hat{v} = (v_1, \dots, v_N) \quad \hat{w} = (w_1, \dots, w_N) \quad v_1 = |z_1| \quad w_1 = v_1 - z_1$$

$$v_i = |z_1| + \sum_{k=1}^{i-1} |z_{k+1} - z_k| \quad w_i = v_i - z_i \quad i = 2, \dots, N$$

allows us to represent the vector $\hat{z}$ by $\hat{z} = \hat{v} - \hat{w}$ with $\hat{v}, \hat{w} \in \widehat{Z} \uparrow$. It is possible, merely by setting

$$\begin{matrix} \xi_1 = v_1 \geq 0 & \xi_{k+1} = v_{k+1} - v_k \geq 0 \\ \eta_1 = w_1 \geq 0 & \eta_{k+1} = w_{k+1} - w_k \geq 0 \end{matrix} \quad k = 1, \dots, N-1$$

which must satisfy the equalities

$$z_k = \sum_{l=1}^{k} (\xi_l - \eta_l) \qquad k = 1, \dots, N \tag{3}$$

In the new variables, functional (2) takes the form

$$\widehat{M}^\alpha[\hat{\xi}, \hat{\eta}] = \alpha \sum_{l=1}^{N} |\xi_l - \eta_l| + \Phi(\hat{\xi} - \hat{\eta})$$

where $\hat{\xi} = (\xi_1, \dots, \xi_N) \in \mathbf{R}_+^N$, $\hat{\eta} = (\eta_1, \dots, \eta_N) \in \mathbf{R}_+^N$ and a new functional Φ is specified, in view of (3), by the equality $\Phi(\hat{\xi} - \hat{\eta}) = \Phi_\alpha(\hat{z})$. In so doing, problem (1) reduces to the finite-dimensional minimization problem with restrictions: find elements $\hat{\xi}^*, \hat{\eta}^* \in \mathbf{R}_+^N$ such that:

$$\widehat{M}^\alpha[\hat{\xi}^*, \hat{\eta}^*] = \inf \{ \widehat{M}^\alpha[\hat{\xi}, \hat{\eta}] : \hat{\xi}, \hat{\eta} \in \mathbf{R}_+^N \} \tag{4}$$

In turn, the problem (4) needs reducing to a form in which we are willing to construct vectors $\hat{a}^*, \hat{b}^* \in \mathbf{R}_+^N$ such that

$$\widehat{L}^\alpha[\hat{a}^*, \hat{b}^*] = \inf \{ \widehat{L}^\alpha[\hat{a}, \hat{b}] : \hat{a}, \hat{b} \in \mathbf{R}_+^N \} \tag{5}$$

Here the functional $\widehat{L}^\alpha[\hat{a}, \hat{b}]$ is as follows:

$$\widehat{L}^\alpha[\hat{a}, \hat{b}] = \alpha \sum_{k=1}^{N} (a_k + b_k) + \Phi(\hat{a} - \hat{b}) \equiv \alpha \widehat{\Omega}(\hat{a}, \hat{b}) + \Phi(\hat{a} - \hat{b})$$

The reduction of such a kind is the usual practice in the **least modulus method** (see, for example, Mudrov and Kushko (1983), Zhukhovitskiĭ and Avdeeva (1964)). In the following theorem we establish an interrelation between the solutions of problem (4) and those of problem (5).

Theorem 1 *If* $(\hat{a}^*, \hat{b}^*)$ *is a solution of problem* (5), *then the vector* $(\hat{\xi}^*, \hat{\eta}^*) = (\hat{a}^*, \hat{b}^*)$ *is a solution of problem* (4).

Proof. Observe that any solution of problem (5) has a characteristic feature lying in the fact that for each $k = 1, \dots, N$ the numbers a_k^* and b_k^* cannot be positive simultaneously. Indeed, when this is not the case, that is, $a_{k_0}^* > 0$ and $b_{k_0}^* > 0$ for $k = k_0$, one can choose a positive number $\lambda \equiv \min(a_{k_0}^*, b_{k_0}^*)$ and define elements $\hat{a}^{**}, \hat{b}^{**} \in \mathbf{R}_+^N$ as follows: $a_k^{**} = a_k^*$, $b_k^{**} = b_k^*$ for $k \neq k_0$ and $a_{k_0}^{**} = a_{k_0}^* - \lambda$, $b_{k_0}^{**} = b_{k_0}^* - \lambda$. Whence it follows that

$$\widehat{L}^\alpha[\hat{a}^{**}, \hat{b}^{**}] = \widehat{L}^\alpha[\hat{a}^*, \hat{b}^*] - 2\alpha\lambda < \widehat{L}^\alpha[\hat{a}^*, \hat{b}^*]$$

which disagrees with the extremality of the element $(\hat{a}^*, \hat{b}^*)$ as a solution of problem (5).

By virtue of the characteristic property of the solutions to problem (5) just established we derive the equality

$$\sum_{k=1}^{N}(a_k^* + b_k^*) = \sum_{k=1}^{N}|a_k^* - b_k^*| \tag{6}$$

yielding $\widehat{L}^\alpha[\hat{a}^*, \hat{b}^*] = \widehat{M}^\alpha[\hat{a}^*, \hat{b}^*]$. If $(\hat{a}^*, \hat{b}^*)$ is none of the solutions to problem (4), then there exist vectors $\hat{\xi}, \hat{\eta} \in \mathbf{R}_+^N$ such that $\widehat{M}^\alpha[\hat{a}^*, \hat{b}^*] > \widehat{M}^\alpha[\hat{\xi}, \hat{\eta}]$. Further, we are concerned with vectors $\hat{\xi}^{**}, \hat{\eta}^{**} \in \mathbf{R}_+^N$ defined by the relations

$$\xi_k^{**} = \begin{cases} \xi_k - \eta_k & \xi_k > \eta_k \\ 0 & \xi_k \leq \eta_k \end{cases}$$

$$\eta_k^{**} = \begin{cases} 0 & \xi_k > \eta_k \\ \eta_k - \xi_k & \xi_k \leq \eta_k \end{cases}$$

Bearing in mind the detailed forms of the functionals $\widehat{M}^\alpha$ and $\widehat{L}^\alpha$, we arrive at the relations

$$\widehat{M}^\alpha[\hat{a}^*, \hat{b}^*] = \widehat{L}^\alpha[\hat{a}^*, \hat{b}^*] > \widehat{M}^\alpha[\hat{\xi}, \hat{\eta}] = \widehat{L}^\alpha[\hat{\xi}^{**}, \hat{\eta}^{**}]$$

But the inequality obtained contradicts the extremality of the quantity $(\hat{a}^*, \hat{b}^*)$ as a solution of problem (5). This provides reason enough to conclude that for any $\hat{\xi}, \hat{\eta} \in \mathbf{R}_+^N$

$$\widehat{M}^\alpha[\hat{a}^*, \hat{b}^*] \leq \widehat{M}^\alpha[\hat{\xi}, \hat{\eta}]$$

and the theorem is completely proved. □

From such reasoning it seems clear (Theorem 1) that for finding an extremal of problem (1) it suffices to solve problem (5) and replace the corresponding variables by formula (3).

Theorem 2 *Under the assumptions of Section* 2 *imposed on the finite-dimensional quantities* $\bar{P}_N$, $\bar{Q}_M$, Ω_1, $\hat{A}_H$ *and the function* $f(x)$, *problem* (5) *is solvable.*

Proof. Indeed, by the conditions of the theorem, the functional $\Phi(\hat{a} - \hat{b})$ is continuous on the set $\mathbf{R}_+^N \times \mathbf{R}_+^N$ as well as the functional $\widehat{\Omega}(\hat{a}, \hat{b})$. Moreover, the functional $\widehat{\Omega}(\hat{a}, \hat{b})$ possesses the compactness property of sets $\widehat{\Omega}_C = \{\hat{a}, \hat{b} \in \mathbf{R}_+^N : \widehat{\Omega}(\hat{a}, \hat{b}) \leq C\}$ $(C \geq 0)$ in $\mathbf{R}_+^N \times \mathbf{R}_+^N$ that owes a debt to the estimate $0 \leq a_k + b_k \leq \widehat{\Omega}(\hat{a}, \hat{b})$, $a_k \geq 0, b_k \geq 0$. This provides support for the view that problem (5) is solvable in complete agreement with Theorem 2.4.2 in the general setting. □

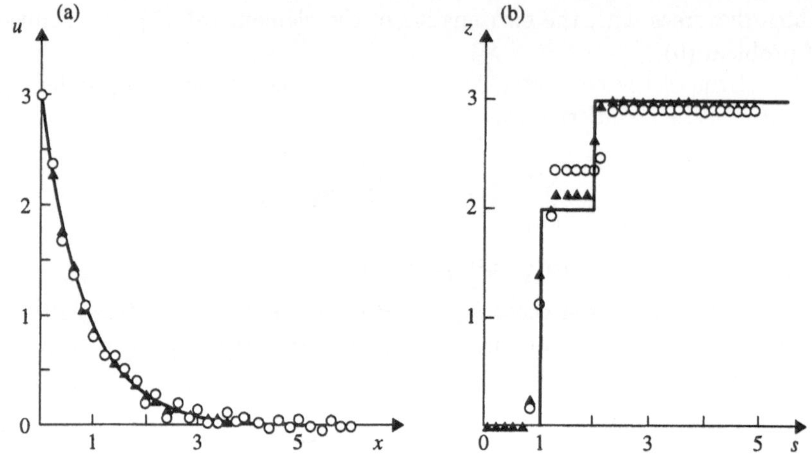

Figure 5.1 Piece-uniform regularization of equation (3.2) with the kernel $K(x,s) = e^{-xs}$ and the right-hand side $\bar{u} = 2e^{-x} + e^{-2x}$ by means of the g.p.d. algorithm: (a) the exact and approximate right-hand sides; (b) the exact and approximate solutions. The exact quantities are depicted by continuous curves, and the approximate quantities are illustrated by symbols ∘∘∘ for $\sigma = 10^{-2}$ and by symbols ▲▲▲ for $\sigma = 10^{-3}$.

In many cases, problem (5) is much simpler than problem (1). This is especially true for the linear operator equations (2.1) when problem (5) falls within the category of problems of quadratic programming on the set of all vectors with nonnegative components. Effective finite-step numerical methods for solving similar problems have been developed (see e.g. Pshenichny and Danilin (1975)).

In dealing with the **g.p.d.**, **g.p.q.** and **g.p.f.** algorithms being used for piece-uniform regularization of equation (2.1) various questions connected with their numerical realization can be fitted into the framework of Section 2.14 in Chapter 2. Adopting its arguments, we are to look for only one solution $\hat{z}^\alpha$ of problem (1) for arbitrary $\alpha > 0$ by solving problem (5) as before.

In conclusion, we give several examples of model piece-uniform regularization of normal solutions to an equation of the form (3.2) in the framework of Sections 2–3. In these examples the minimization is implemented for the smoothing functional (1) with $p = 0$ by using the procedure of the present subsection according to which, first, we must solve problem (5) and then find from its solution $\hat{\xi}^\alpha = \hat{a}^\alpha$ and $\hat{\eta}^\alpha = \hat{b}^\alpha$ by formula (3) an element $\hat{z}^\alpha$ minimizing the indicated smoothing functional over the space $\mathbf{R}^N$.

The results of some computations are given in Figs. 5.1–5.3. They represent the exact quantities $\bar{u}$ and $\bar{z}$, the vector of the approximate right-

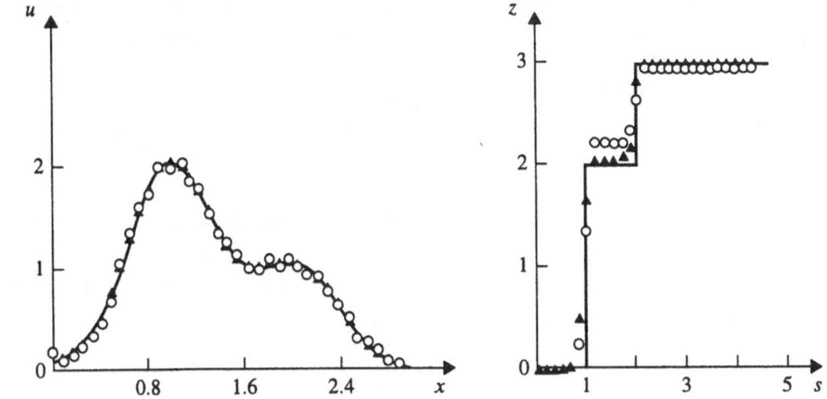

Figure 5.2 Piece-uniform regularization of equation (3.2) with the kernel $K(x,s) = \exp\{-4(x-s)^2\}$ and the right-hand side $\bar{u} = 2\exp\{-4(x-1)^2\} + \exp\{-4(x-2)^2\}$.

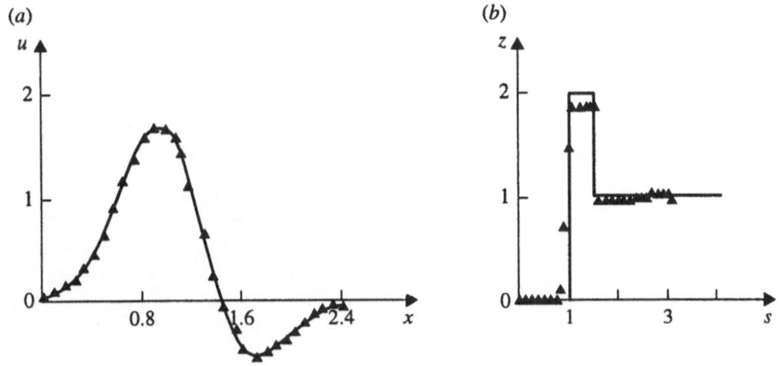

Figure 5.3 Piece-uniform regularization of equation (3.2) with the kernel $K(x,s) = \exp\{-4(x-s)^2\}$ and the right-hand side $\bar{u} = 2\exp\{-4(x-1)^2\} - \exp\{-4(x-3/2)^2\}$.

hand side $\hat{u}_M$ of equation (3.2) as well as the vector $\hat{z}^{\alpha_\eta}$ obtained by means of the algorithm of piece-uniform regularization (**f.d.g.p.d.**) at various perturbation levels σ of the right-hand side of the equation in hand.

Fig. 5.1 shows the solution to equation (3.2) with the kernel $K(x,s) = \exp\{-xs\}$. Fig. 5.2 and 5.3 correspond to the same equation but with the kernel $K(x,s) = \exp\{-4(x-s)^2\}$.

As a matter of experience, extra sensitivity of approximate solutions to the errors in the right-hand side of equation (3.2) is well-known to scientists (see Lanczos (1956)). From Fig. 5.1 this fact is almost obvious.

Remark Numerical methods for solving ill-posed problems on various sets of monotone, convex and some other functions are outlined in Goncharskiĭ, Gushchina *et al.* (1972), Goncharskiĭ and Stepanov (1980), Tikhonov *et al.* (1983, 1995). Fortran programs for the implementation of these methods are available in Tikhonov *et al.* (1983, 1995). The results similar to those of Sections 5.1–5.2 and 5.4 in Chapter 5 can be obtained for the functions of several variables with bounded variation in the sense of Vitali and Hardy (see, for example, Clarkson and Adams (1933)).

CHAPTER 6

Applications to solving linear algebraic problems

In this chapter, applications of the **g.p.d.** algorithm to the stable solution of linear algebraic problems are given special investigation. Most of the propositions are of independent value although, for the present volume, they are used only as part of the auxiliary mathematical apparatus. Some of them were known earlier and the rest were discovered and proven in recent years in connection with development of the method with optimal accuracy order.

6.1 The fundamental problem

In trying to produce several stable methods for solving systems of linear algebraic equations (SLAE) having the form

$$\bar{A}z = \bar{u} \qquad z \in \mathbf{R}^n \qquad \bar{u} \in \mathbf{R}^m \tag{1}$$

we are dealing with a nonzero real matrix $\bar{A}$ of order $m \times n$ and $\bar{u} \neq 0$. This system may be unsolvable in the classical sense. However, it can always have a nonempty set of pseudosolutions Z^* consisting of all vectors $z^* \in \mathbf{R}^n$ for which

$$\|\bar{A}z^* - \bar{u}\| = \inf\left\{\|\bar{A}z - \bar{u}\|: z \in \mathbf{R}^n\right\}$$

Throughout this chapter, all the norms are supposed to be Euclidean. For any compatible system (1) the pseudosolutions so defined are identical with ordinary solutions themselves.

It is well-known that for SLAE (1) there exists a unique normal pseudosolution $\bar{z}$, that is, a pseudosolution for which

$$\|\bar{z}\| = \inf\{\|z\|: z \in Z^*\}$$

For the solvable system (1) normal pseudosolution is called a **normal solution** (see Tikhonov (1965 e), Tikhonov and Arsenin (1977)). In what follows we will assume $\bar{z} \neq 0$.

Suppose that instead of the exact data $(\bar{A}, \bar{u})$ of system (1) we are given their approximations (A_h, u_σ), where A_h is the approximate $(m \times n)$-matrix with

$$\|A_h - \bar{A}\| \leq h$$

and u_σ the approximate right-hand side of the system with

$$\|u_\sigma - \bar{u}\| \leq \sigma$$

It is easy to verify that in the general case the problem of finding a normal pseudosolution becomes unstable with respect to perturbations of the matrix and can depend on the range of situations to be considered. In order to understand the nature of this a little better, consider the following example.

Example 1 Let

$$\bar{A} = \begin{pmatrix} 1 & 0 \\ 0 & 0 \end{pmatrix} \qquad \bar{u} = \begin{pmatrix} 1 \\ 1 \end{pmatrix} \qquad A_h = \begin{pmatrix} 1 & 0 \\ 0 & h \end{pmatrix}$$

$$u_\sigma = \bar{u} \quad \sigma = 0 \quad z = \begin{pmatrix} x \\ y \end{pmatrix}$$

Direct calculations by the definition of the normal pseudosolution of system (1) give $\bar{z} = (1,0)^{\mathrm{T}}$. The system $A_h z = u_\sigma$ with perturbed data has the unique solution $\bar{z}_h = (1, 1/h)^{\mathrm{T}}$. Because of its form the quantities $\bar{z}_h$ do not converge to $\bar{z}$ as $h \to 0$, thus causing instability of the normal pseudosolution of the problem at hand.

The fundamental problem of this chapter consists of constructing from the available approximate data $(A_h, u_\sigma, h, \sigma)$ a stable approximation z_δ, $\delta \equiv (h, \sigma)$, to normal pseudosolution $\bar{z}$ of system (1) over $\mathbf{R}^n$:

$$\|z_\delta - \bar{z}\| \to 0 \qquad \text{as} \qquad \delta \to 0$$

In principle, such a statement of the problem represents merely a particular case of the general setting of Section 3.1 in Chapter 3 with $Z = D = \mathbf{R}^n$, $U = \mathbf{R}^m$, $\Omega(z) = \|z\|^2$, $\psi(\delta, \Omega) = \sigma + h\|z\|$ and $\bar{Z} = \{\bar{z}\}$. Therefore, the algorithms developed in Chapter 3 may be of help in solving the fundamental problem posed above. What has become a very conventional view is that problem (1) is linear and finite-dimensional. For this reason alone, the algorithms which have been under consideration acquire more properties and peculiarities. To reveal some of them, the algorithm for finding a normal pseudosolution $\bar{z}$ by means of the **generalized principle of discrepancy (g.p.d.)** is considered first.

The **g.p.d.** algorithm for solving the fundamental problem is based on a smoothing functional of the form

$$M^\alpha[z] = \alpha\|z\|^2 + \|A_h z - u_\sigma\|^2 \qquad z \in \mathbf{R}^n \qquad \alpha > 0 \tag{2}$$

By regarding δ to be fixed and $\alpha > 0$ to be given, we state that the problem of minimizing a smoothing functional, in which it is necessary to obtain an element z^α such that

$$M^\alpha[z^\alpha] = \inf\left\{M^\alpha[z] \colon z \in \mathbf{R}^n\right\} \tag{3}$$

possesses a unique solution. This fact follows either from Example 4 of Section 1.5 in Chapter 1 or from Lemma 2.11.1, since any such smoothing functional is the sum of the strongly convex in $\mathbf{R}^n$ functional $\|z\|^2$ and the convex functional of discrepancy $\|A_h z - u_\sigma\|^2$. Moreover, a necessary and sufficient condition for z^α to be a solution of the extremal problem (3) over $\mathbf{R}^n$ with strongly convex and continuously differentiable functional (2) is provided by $\text{grad} M^\alpha[z^\alpha] = 0$, which is equivalent to the equation

$$\alpha z^\alpha + A_h^{\mathrm{T}} A_h z^\alpha = A_h^{\mathrm{T}} u_\sigma \tag{4}$$

(see, for example, Tikhonov and Arsenin (1977) or Example 1.5.4). Such operator equations and relevant systems of linear algebraic equations for finding z^α have been studied in detail and effectively solved earlier.

Our starting point is the definition of the generalized discrepancy

$$\rho(\alpha) = \|A_h z^\alpha - u_\sigma\|^2 - \left(\lambda_\delta + \sigma + h\|z^\alpha\|\right)^2 \qquad \alpha > 0 \tag{5}$$

whose use permits us to choose properly the regularization parameter. Here the number

$$\lambda_\delta = \inf\left\{\|A_h z - u_\sigma\| + \sigma + h\|z\| \colon z \in \mathbf{R}^n\right\}$$

is the generalized incompatibility measure of SLAE (1) first introduced in Kochikov *et al.* (1984) and Wentsel' *et al.* (1984). Due to the uniqueness of the solution to problem (3) the generalized discrepancy is continuous for $\alpha > 0$ (see Section 2.6 in Chapter 2). It is worth noting here that its limiting values $\rho_0 = \rho(+0)$ and $\rho_\infty = \rho(+\infty)$ are subject to the inequalities $\rho_0 < 0$ and $\rho_\infty > 0$ for all sufficiently small σ and h. This property follows from Theorem 2.7.3, since the approximation measure $\Psi(\delta, \Omega) = \sigma + h\Omega^{1/2}$, $h \neq 0$, is a monotonically increasing function with respect to the second argument and the condition $\bar{z} \neq 0$ is satisfied. Retaining the notation of Chapter 2 we interpret the preceding condition as meaning that we should have $Z_0 \cap Z^* = \varnothing$.

Theorem 2.11.1 implies that the equation $\rho(\alpha) = 0$, which is responsible for the proper choice of the regularization parameter α_δ in the **g.p.d.** algorithm, has a unique solution $\alpha_\delta > 0$. Observe that the function $\rho(\alpha)$ defined for $\alpha > 0$ is monotonically increasing (see Lemma 3.4.2).

Thus, if $\delta = (h, \sigma)$ is held fixed, the **g.p.d.** algorithm gives the unique approximate solution $z_\delta \equiv z^{\alpha_\delta}$ of system (1) that converges by Theorem 2.7.5 to $\bar{z}$ as $\delta \to 0$.

When more information about the system (1) compatibility is prescribed, we may put $\lambda_\delta = 0$ in (5) (see Section 3.3 in Chapter 3).

The **g.p.d.** algorithm for solving the fundamental problem is based on the data $(A_h, u_\sigma, h, \sigma)$. However, many scientists prefer its modification with no use of the σ-error in the explicit form (see Leonov (1985 b)).

For the reader's convenience we introduce the normed spaces of all matrices of orders $m \times n$, $n \times m$, $m \times m$ and $n \times n$ with Euclidean norms $\|\cdot\|$,

$\|\cdot\|_*$, $\|\cdot\|_m$ and $\|\cdot\|_n$ and denote them by $\mathfrak{A}$, $\mathfrak{A}^*$, $\mathfrak{A}_m$ and $\mathfrak{A}_n$, respectively. It is well-known that any normal pseudosolution $\bar{z}$ of problem (1) can be expressed in terms of the exact data $(\bar{A}, \bar{u})$ by $\bar{z} = \bar{A}^+\bar{u}$, where the pseudo inverse $\bar{A}^+$ is just a solution of the extremal problem in which it is necessary to find a matrix $\widetilde{Z} \in \mathfrak{A}^*$ such that

$$\|\bar{A}\widetilde{Z} - E\|_m = \inf\left\{\|\bar{A}Z - E\|_m \colon Z \in \mathfrak{A}^*\right\} \tag{6}$$

Here $E \in \mathfrak{A}_m$ is the unit matrix. This problem may be nonuniquely solvable. It is easy to verify that along with $\widetilde{Z} = \bar{A}^+$ the extremal problem of interest admits solutions of the form $\widetilde{Z} = \bar{A}^+ + q(I - \bar{A}^+\bar{A})$, where I is the unit matrix from $\mathfrak{A}_n$, q is an arbitrary constant, $\bar{A}^+$ is a unique normal solution of problem (6). Of course, it is to be understood that if $\mathfrak{A}_0^*$ is the set of solutions to problem (6), then

$$\|\bar{A}^+\|_* = \inf\left\{\|\widetilde{Z}\|_* \colon \widetilde{Z} \in \mathfrak{A}_0^*\right\} \tag{7}$$

Thus, the extremal problem (6)–(7) of obtaining the pseudoinverse $\bar{A}^+$ is a particular case of the problems considered in Chapter 2. The availability of the generalized discrepancy principle for extremal problems (see Section 2.7 in Chapter 2) has had a significant impact on constructions of regularized approximations to the pseudoinverse of interest. A smoothing functional of the form

$$M^\alpha[Z] = \alpha\|Z\|_*^2 + \|A_h Z - E\|_m^2 \qquad \alpha > 0 \qquad Z \in \mathfrak{A}^* \tag{8}$$

can serve as a necessary background for the generalized principle of discrepancy capable of elucidating many facets of algebraic problems. For each $\alpha > 0$ such a functional has the unique extremal

$$Z^\alpha = (\alpha I + A_h^{\mathrm{T}} A_h)^{-1} A_h^{\mathrm{T}}$$

realizing the greatest lower bound in $\mathfrak{A}^*$. This provides reason enough to refer to the generalized discrepancy

$$\rho(\alpha) = \|A_h Z^\alpha - E\|_m^2 - (\Lambda_h + h\|Z^\alpha\|_*)^2$$

with

$$\Lambda_h \equiv \inf\left\{\|A_h Z - E\|_m + h\|Z\|_* \colon Z \in \mathfrak{A}^*\right\}$$

which is a continuous monotonically increasing function defined for $\alpha > 0$.

The key results of Section 2.7 in Chapter 2 entail that the equation $\rho(\alpha) = 0$ possesses a unique solution $\alpha(h) > 0$ if the error $h > 0$ is small enough. If we regard the matrix $Z^{\alpha(h)}$ as an approximation to $\bar{A}^+$ then, by Theorem 2.7.5, the convergence $Z^{\alpha(h)} \to \bar{A}^+$ occurs as $h \to 0$.

There are other approaches for producing regularized approximations to $\bar{A}^+$ (see, e.g. Meleshko and Zadachin (1987) and Morozov (1984)).

The regularized approximations $Z^{\alpha(h)}$ can be used to construct approximations to $\bar{z}$: $z_\delta = Z^{\alpha(h)} u_\sigma$. Then the inequality

$$\|z_\delta - \bar{z}\| \le \|Z^{\alpha(h)} - \bar{A}^+\|_* \cdot \|u_\sigma\| + \|\bar{A}^+\|_* \cdot \|u_\sigma - \bar{u}\|$$

provides the convergence $z_\delta \to \bar{z}$ as $\delta = (h, \sigma) \to 0$.

The above algorithm for finding approximations to a normal pseudosolution of SLAE (1) does not depend on σ, but relies essentially on h. This is the case due to the specific character of problem (1) lying in the normal solvability of the operator $\bar{A}$. The results of Bakushinskiĭ (1984) and Section 3.7 in Chapter 3 demonstrate that it is impossible to develop any regularizing algorithm for solving SLAEs of the general form (1) without the need for the error h.

The principle difference between compatible and incompatible SLAEs (1) is revealed in the algorithms for the stable solution of SLAE (1) with the use of the smoothing functional (2) (or (8)) (and, in particular, in the **g.p.d** algorithm). As stated in Vaĭnikko and Veretennikov (1986) as well as in Gilyazov and Morozov (1984), the compatibility is significant in the attainment of the optimal accuracy order of an approximate solution.

Before taking up the general case, we begin by considering the example of an incompatible system along with the data $(\bar{A}, \bar{u}, A_h, u_\sigma)$ taken from Example 1. In this case we may attempt functional (2) in the form

$$M^\alpha[z] = M^\alpha(x, y) = \alpha(x^2 + y^2) + (x-1)^2 + (hy-1)^2$$

Having solved problem (3) we obtain

$$z^\alpha = (x^\alpha, y^\alpha)^{\mathrm{T}} \qquad x^\alpha = 1/(1+\alpha) \qquad y^\alpha = h/(\alpha + h^2)$$

whence it follows that

$$\|z^\alpha - \bar{z}\|^2 = \alpha^2/(1+\alpha)^2 + h^2/(\alpha+h^2)^2 \equiv p^2(\alpha)$$

It is not difficult to verify that $\min p(\alpha) \asymp \sqrt{h}$. Thus, in the given example, for any choice of the regularization parameter, $\sqrt{h}$ is the best possible accuracy order of the approximate solution and cannot be improved.

On the other hand, it turns out that for the compatible SLAE (1) the following theorem is valid.

Theorem 1 *When a normal solution $\bar{z}$ of the compatible system* (1) *is sought by the regularization method with the use of the smoothing functional* (2), *the regularization parameter $\alpha = \alpha(h)$ can be chosen in such a way to satisfy, for $0 \le h < h_0 = const$ and any $\sigma \ge 0$, the relation*

$$\|z^{\alpha(h)} - \bar{z}\| \le \frac{\|\bar{A}^+\|_*[\sigma + h(1+\sqrt{2})\|\bar{z}\|]}{1 - h\|\bar{A}^+\|_*} = O(h+\sigma) \tag{9}$$

This estimate is unimprovable in the order of its ingredients $\sigma, h, \|\bar{A}^+\|_*$.

The key result of this section has been established by Vaĭnikko and Veretennikov (1986). The accuracy order of estimate (9) was also studied by Gilyazov and Morozov (Morozov and Gilyazov 1979, Gilyazov and Morozov 1984).

Exploiting these facts we conclude that for solvable SLAEs the regularization method with a proper choice of the regularization parameter α

quarantees the optimal accuracy order of an approximate solution, namely of $O(h+\sigma)$. For more detail on this point, we recommend the reader Section 5 of this chapter.

Returning to the **g.p.d.** algorithm it is worth emphasizing one thing. The above example demonstrates that the **g.p.d.** for incompatible SLAEs does not guarantee the optimal accuracy order. We now show that for the compatible system (1) the accuracy order of the approximate solution obtained by means of the **g.p.d.** algorithm will be optimal. With this aim, we employ the **m.g.p.d.** algorithm according to which a unique root of the equation with a monotonically increasing function

$$\rho(\alpha) = \|A_h z^\alpha - u_\sigma\| - C(\sigma + h\|z^\alpha\|) = 0 \qquad C > 1 \tag{10}$$

can be adopted as the needed regularization parameter α_δ (cf. (5)). The possibility of such a choice of the regularization parameter and the convergence $z^{\alpha_\delta} \to \bar{z}$ as $\delta \to 0$ have been justified in Section 3.3 in Chapter 10. By the same token, $\|z^{\alpha_\delta}\| \leq \|\bar{z}\|$.

Theorem 2 *If $\bar{u} \in Im\bar{A}$, that is, SLAE (1) is solvable, then the* **m.g.p.d.** *algorithm with recovering α_δ from equation (10) is of optimal accuracy order.*

To prove Theorem 3 we introduce two auxiliary lemmas.

Lemma 1 (Vaĭnikko and Veretennikov (1986)) *If $\bar{u} \in Im\bar{A}$, $b = const > \|\bar{A}^+\|_* / (\|\bar{A}^+\|_* - h)$ and a number $\alpha(\delta)$ satisfies the equation*

$$\beta(\alpha) \equiv \|A_h z^\alpha - u_\sigma\| = b(\sigma + h\|\bar{z}\|)$$

then $\|z^{\alpha(\delta)} - \bar{z}\| = O(\sigma + h)$

Lemma 2 *For $0 < \alpha_1 \leq \alpha \leq \alpha_2$ the inequality is valid:*

$$\|z^{\alpha_2} - z^\alpha\| \leq \|z^{\alpha_2} - z^{\alpha_1}\|$$

Proof. Via the singular value decomposition of the matrix A_h given by $A_h = URV^{\mathrm{T}}$ (see Section 1.6 in Chapter 1) one can find from equation (4) the element

$$z^\alpha = V(\alpha I + R^{\mathrm{T}}R)^{-1}R^{\mathrm{T}}U^{\mathrm{T}}u_\sigma$$

Then, denoting $w \equiv U^{\mathrm{T}}u_\sigma$ and taking into account the orthogonality of the matrix V we arrive at the chain of relations

$$\begin{aligned}\|z^{\alpha_2} - z^\alpha\|^2 &= \|V[(\alpha_2 I + R^{\mathrm{T}}R)^{-1}R^{\mathrm{T}}w \\ &\quad -(\alpha I + R^{\mathrm{T}}R)^{-1}R^{\mathrm{T}}w]\|^2 \\ &= \|(\alpha_2 I + R^{\mathrm{T}}R)^{-1}(\alpha I + R^{\mathrm{T}}R)^{-1} \\ &\quad \times(\alpha I + R^{\mathrm{T}}R - \alpha_2 I - R^{\mathrm{T}}R)R^{\mathrm{T}}w\|^2 \\ &= (\alpha - \alpha_2)^2\|(\alpha_2 I + R^{\mathrm{T}}R)^{-1} \\ &\quad \times(\alpha I + R^{\mathrm{T}}R)^{-1}R^{\mathrm{T}}w\|^2\end{aligned}$$

Rewriting the last equality in an expanded form and involving $R = \mathrm{diag}(\rho_1, \dots, \rho_M) \in \mathfrak{A}$, $M \equiv \min(m,n)$ and $w = (w_1, \dots, w_m)$ we obtain the estimate

$$\begin{aligned} \|z^{\alpha_2} - z^{\alpha}\|^2 &= (\alpha_2 - \alpha)^2 \sum_{k=1}^{M} \frac{\rho_k^2 w_k^2}{(\alpha_2 + \rho_k^2)^2 (\alpha + \rho_k^2)^2} \\ &\leq (\alpha_2 - \alpha_1)^2 \sum_{k=1}^{M} \frac{\rho_k^2 w_k^2}{(\alpha_2 + \rho_k^2)^2 (\alpha_1 + \rho_k^2)^2} \\ &= \|z^{\alpha_2} - z^{\alpha_1}\| \end{aligned}$$

thereby completing the proof of the lemma. □

Proof of Theorem 2. Let a number $\varepsilon (0 < \varepsilon < 1)$ be such that $C(1-\varepsilon) > 1$. We choose small numbers σ_0 and h_0 in such a way to satisfy for $0 \leq \sigma < \sigma_0$ and $0 \leq h < h_0$ the relations

$$\|z^{\alpha_\delta}\| \geq (1-\varepsilon)\|\bar{z}\| \qquad C(1-\varepsilon) > \frac{\|\bar{A}^+\|_*}{\|\bar{A}^+\|_* - h}$$

Such numbers σ_0 and h_0 do exist, otherwise by means of a convergent sequence $\delta_n = (\sigma_n, h_n) \to 0$ as $n \to \infty$ the contrary should be explicitly stated:

$$\|z^{\alpha_{\delta_n}}\| < (1-\varepsilon)\|\bar{z}\| \qquad C(1-\varepsilon) \leq \frac{\|\bar{A}^+\|_*}{\|\bar{A}^+\|_* - h_n}$$

Passing in the above inequalities to the limit as $n \to \infty$ and taking into account the convergence $z^{\alpha_{\delta_n}} \to \bar{z}$, we are led to the relations $\|\bar{z}\| \leq (1-\varepsilon)\|\bar{z}\|$ and $C(1-\varepsilon) \leq 1$. But this contradicts our assumption concerning the number ε and provides support for imposing the constraints $0 \leq \sigma < \sigma_0$ and $0 \leq h < h_0$. Putting these together with the inequality $\|z^{\alpha_\delta}\| \leq \|\bar{z}\|$, we obtain

$$\begin{aligned} C(1-\varepsilon)(\sigma + h\|\bar{z}\|) &\leq C(\sigma + h(1-\varepsilon)\|\bar{z}\|) \\ &\leq C(\sigma + h\|z^{\alpha_\delta}\|) \\ &\leq C(\sigma + h\|\bar{z}\|) \end{aligned} \qquad (*)$$

When $\alpha_1(\delta)$ and $\alpha_2(\delta)$ are recovered from the equations

$$\beta(\alpha) = C(1-\varepsilon)(\sigma + h\|\bar{z}\|) \qquad \beta(\alpha) = C(\sigma + h\|\bar{z}\|)$$

respectively, the relations $0 \leq \alpha_1(\delta) \leq \alpha_\delta \leq \alpha_2(\delta)$ follow from the inequalities $(*)$ because the function $\beta(\alpha)$ is increasing. According to Lemma 2 we establish

$$\|z^{\alpha_2(\delta)} - z^{\alpha_\delta}\| \leq \|z^{\alpha_2(\delta)} - z^{\alpha_1(\delta)}\|$$

yielding, due to Lemma 1,

$$\begin{aligned}\|z^{\alpha_\delta} - \bar{z}\| &\le \|z^{\alpha_\delta} - z^{\alpha_2(\delta)}\| + \|z^{\alpha_2(\delta)} - \bar{z}\| \\ &\le \|z^{\alpha_2(\delta)} - z^{\alpha_1(\delta)}\| + \|z^{\alpha_2(\delta)} - \bar{z}\| \\ &\le 2\|z^{\alpha_2(\delta)} - \bar{z}\| + \|z^{\alpha_1(\delta)} - \bar{z}\| \\ &= O(\sigma + h)\end{aligned}$$

□

Observe that for certain incompatible systems (1), namely provided that the condition of the pseudosolution uniqueness holds, the optimal accuracy order of the **g.p.d.** algorithm can be justified by the results of Sections 2.13 in Chapter 2 and 3.6 in Chapter 3.

In accordance with what has been said it is necessary to refine the fundamental problem by rearranging its statement. At present, we are only interested in a stable method for the approximate determination of a normal pseudosolution of system (1) enabling us to reach the optimal accuracy order disregarding to the fact whether the system is compatible or not.

Such a method has been developed in recent years on the basis of the **g.p.d. (g.m.d.)** algorithm by a special choice of the functional Ω. The method from the works of Leonov (1987 a, 1991) that possesses a number of optimal properties is given special investigation in subsequent sections, placing particular emphasis on questions of efficiency and accuracy.

6.2 Minimal pseudoinverse matrix (m.p.m.) method

Consider SLAE (1.1) and assume that the matrix $\bar{A}$ belongs to a set $\mathfrak{A}_0$, which is closed in the space $\mathfrak{A}$. In particular, we might agree with $\mathfrak{A}_0 = \mathfrak{A}$. The inclusion $\bar{A} \in \mathfrak{A}_0$ falls in the category of the possible *a priori* information on the special structure of the matrix $\bar{A}$ (if any).

Given the approximate data (A_h, h) of problem (1.1), for later use we define, holding h fixed, the set of matrices

$$\mathfrak{A}_h \equiv \{A \in \mathfrak{A}_0 \colon \|A - A_h\| \le h\}$$

The **minimal pseudoinverse matrix (m.p.m.) method** for solving the fundamental problem is connected with the following extremal problem in which it is necessary to obtain a matrix $\widetilde{A}_h \in \mathfrak{A}_h$ such that

$$\|\widetilde{A}_h^+\|_* = \inf\{\|A^+\|_* \colon A \in \mathfrak{A}_h\} \equiv \varkappa_h \tag{1}$$

Any solution $\widetilde{A}_h$ of problem (1) is called a **matrix of the m.p.m. method** and the associated pseudoinverse $\widetilde{A}_h^+$ is called the **minimal pseudoinverse matrix** for the data (A_h, h).

The construction of the element $z_\eta \equiv \widetilde{A}_h^+ u_\sigma$ $(\eta \equiv (h, \sigma))$ necessitates approximating a normal pseudosolution $\bar{z}$ of SLAE (1.1) by means of the **m.p.m.** method.

In this context, it is of importance to analyze a possibility of numerical realization of this method and to establish its properties. The following result provides an excellent start in this situation.

Lemma 1 *Let a sequence of matrices* $\{C_n\} \subset \mathfrak{A}$ *converge to* $C_0 \in \mathfrak{A}$ *as* $n \to \infty$ *and*

$\|C_n^+\|_* \le K = \text{const}$ for any n, that is, their norms be uniformly bounded. Then the convergence $C_n^+ \to C_0^+$ takes place as $n \to \infty$.

Proof. Due to the uniform boundedness of norms $\|C_n^+\|_*$ in the finite-dimensional space of matrices $\mathfrak{A}^*$, the set $\{C_n^+\}$ is compact in $\mathfrak{A}^*$. Therefore, the sequence $\{C_n^+\}$ contains a convergent subsequence, say $\{C_{n_l}^+\}$, it being understood that $C_{n_l}^+ \to D_0 \in \mathfrak{A}^*$ as $l \to \infty$. We now show that its limit D_0 coincides with C_0^+. For this, we proceed as usual. This amounts to choosing an arbitrary element $u \in \mathbf{R}^m$ and constructing a sequence of normal pseudosolutions $z_l = C_{n_l}^+ u$ of the systems having the form

$$C_{n_l} z = u \tag{2}$$

We learn from Section 1.6 in Chapter 1 that any such element z_l is subject to the equation

$$C_{n_l}^{\mathrm{T}} C_{n_l} z_l = C_{n_l}^{\mathrm{T}} u \tag{3}$$

The convergence of the sequence $\{C_{n_l}^+\}$ to D_0 implies that the elements $z_l = C_{n_l}^+ u$ converge to $z_0 \equiv D_0 u$ as $l \to \infty$. Along with the convergence $C_{n_l} \to C_0$ given by the condition of the lemma, this allows one to deduce from equation (3) as $l \to \infty$ that

$$C_0^{\mathrm{T}} C_0 z_0 = C_0^{\mathrm{T}} u$$

By definition, this means that z_0 is just a pseudosolution of the system $C_0 z = u$.

To decide for yourself whether z_0 is a normal pseudosolution of this system, it suffices to prove that $z_0 \perp \mathrm{Ker} C_0$, that is, the existence of an element $v_0 \in \mathbf{R}^m$ for which $z_0 = C_0^{\mathrm{T}} v_0$ (see Theorem 1.6.1). For normal pseudosolutions z_l of systems having the form (2) there exist elements $v_l \in \mathbf{R}^m$ such that $z_l = C_{n_l}^{\mathrm{T}} v_l$. Of course, the choice of such elements v_l may be nonunique. In particular, we might agree with $v_l = (C_{n_l}^{\mathrm{T}})^+ z_l$. Then the convergences $C_{n_l}^+ \to D_0$ and $z_l \to z_0$ and the equality $(C^{\mathrm{T}})^+ = (C^+)^{\mathrm{T}}$ imply the relations $v_l = (C_{n_l}^{\mathrm{T}})^+ z_l \to D_0^{\mathrm{T}} z_0 \equiv v_0 \in \mathbf{R}^m$. Finally, taking the equalities $C_{n_l}^{\mathrm{T}} v_l = z_l$ to the limit, we obtain $C_0^{\mathrm{T}} v_0 = z_0$, thus demonstrating that z_0 is a normal pseudosolution of the system $C_0 z = u$.

Thus, by means of the matrix D_0 we are able to calculate a normal pseudosolution $z_0 = D_0 u$ of the system $C_0 z = u$ for each $u \in \mathbf{R}^m$. By the definition of pseudoinverse and in view of its uniqueness, one can conclude that $D_0 = C_0^+$ and thereby $C_{n_l}^+ \to C_0^+$ as $l \to \infty$. Since the convergent sequence $\{C_{n_l}^+\}$ is arbitrary, C_0^+ appears to be the unique limit of the sequence $\{C_n^+\}$. □

Corollary 1 *Let $\{C_n\}$ be a sequence from $\mathfrak{A}$, that is, $C_n \in \mathfrak{A}$ for each n. If the estimates $\|C_n\| \le M_1$ and $\|C_n^+\|_* \le M_2$ are valid uniformly in n with positive constants M_1 and M_2, then the set $\{C_n\}$ is compact in the space $\mathfrak{A}$ and the set $\{C_n^+\}$ is compact in the space $\mathfrak{A}^*$.*

The next step is to decide for yourself whether problem (1) is solvable or not. We contrived to do it. The following assertion serves to motivate our answer.

Theorem 1 *Problem* (1) *is solvable for any data (A_h, h) with $A_h \in \mathfrak{A}$ and $h \ge 0$.*

Proof. Let $\{A_N\}$ be an arbitrary minimizing sequence of matrices related to problem (1): $A_N \in \mathfrak{A}_h$ and $\|A_N^+\|_* \to \varkappa_h$ as $N \to \infty$. Then the inequality $\|A_N - A_h\| \le h$ implies the estimate $\|A_N\| \le \|A_h\| + h \equiv M_1$. On the other hand, the convergence $\|A_N^+\|_* \to \varkappa_h$ as $N \to \infty$ causes the boundedness of norms:

$$\|A_N^+\|_* \le M_2$$

According to Corollary 1, the set $\{A_N\} \subset \mathfrak{A}_h$ is compact in the space $\mathfrak{A}$ and, since the set $\mathfrak{A}_h$ is closed in $\mathfrak{A}$, there exists a subsequence $\{A_{N_l}\}$ converging, as $l \to \infty$, to a matrix A_0 from the set $\mathfrak{A}_h$. On the other hand, the convergence $A_{N_l}^+ \to A_0^+$ occurs due to Lemma 1. By the definition of the minimizing sequence $\{A_{N_l}\}$ for problem (1), it follows from the foregoing that

$$\lim_{l\to\infty} \|A_{N_l}^+\|_* = \|A_0^+\|_* = \varkappa_h = \inf\{\|A^+\|_* \colon A \in \mathfrak{A}_h\}$$

Thus, the matrix A_0 gives a solution of problem (1). □

Theorem 2 *Let $\mathfrak{A}_0$ be a subspace of $\mathfrak{A}$. If $\|A_h\| > h$, then all of the solutions $\widetilde{A}_h$ of problem* (1) *satisfy the equality $\|\widetilde{A}_h - A_h\| = h$.*

Proof. Assume to the contrary that the theorem fails to be true and for some solution $\widetilde{A}_h$ of problem (1) the inequality $\|\widetilde{A}_h - A_h\| < h$ appears to be strict. Common practice involves the introduction of the function $f(\lambda) \equiv \|\lambda\widetilde{A}_h - A_h\|$ defined for $\lambda \ge 1$. In accordance with the original hypothesis, $f(1) < h$. On the other hand, the condition of the theorem yields $\|A_h\| > h$. From here it seems clear that the zero matrix is not contained in the set $\mathfrak{A}_h$ and, therefore, the corresponding solution $\widetilde{A}_h$ of problem (1) cannot vanish. This provides reason enough to conclude that

$$f(\lambda) = \lambda\|\widetilde{A}_h - (\lambda^{-1})A_h\| \to +\infty$$

as $\lambda \to +\infty$. Along with continuity of the function $f(\lambda)$ for $\lambda \ge 1$, this limit, relation serves to motivate the existence of a root $\lambda_h > 1$ to the equation

$f(\lambda) = h$. As far as $\mathfrak{A}_0$ is a subspace of $\mathfrak{A}$, we would have $\lambda_h \widetilde{A}_h \in \mathfrak{A}_0$ and $\|\lambda_h \widetilde{A}_h - A_h\| = h$, yielding $\lambda_h \widetilde{A}_h \in \mathfrak{A}_h$. In so doing

$$\|(\lambda_h \widetilde{A}_h)^+\|_* = \lambda_h^{-1} \|\widetilde{A}_h^+\|_* < \|\widetilde{A}_h^+\|_*$$

which disagrees with the preceding conclusion that $\widetilde{A}_h^+$ is a solution of problem (1). The obtained contradiction completes the proof of the theorem. □

It is worth emphasizing here that the condition of the theorem $\|A_h\| > h$ holds true for all sufficiently small h in the case where $\bar{A} \neq 0$.

Corollary 2 *Let the conditions of Theorem* 2 *hold. Then problem* (1) *is equivalent to the following one: we wish to find a matrix* $\widetilde{A}_h \in \mathfrak{A}_0$ *such that*

$$\|\widetilde{A}_h^+\|_* = \inf\left\{\|A^+\|_* \colon A \in \mathfrak{A}_0, \|A - A_h\| = h\right\} \tag{4}$$

Notice that the proof of Theorem 2 remains valid in the case where $\mathfrak{A}_0$ is a cone in $\mathfrak{A}$.

We spoke above about minimal pseudoinverse matrices. It seems worthwhile giving simple examples to help to motivate what is done. Later, we will elaborate on this for rather complicated cases.

Example 1 Let

$$\bar{A} = \begin{pmatrix} 1 & 0 \\ 0 & 0 \end{pmatrix} \qquad A_h = \begin{pmatrix} 1 & 0 \\ 0 & H \end{pmatrix} \qquad 0 < H < h$$

$$\mathfrak{A}_0 \equiv \left\{ \begin{pmatrix} 1 & 0 \\ 0 & x \end{pmatrix} : x \in \mathbf{R} \right\} \subset \mathfrak{A} \equiv \mathfrak{A}_2$$

For any matrices from $\mathfrak{A}_0$ it is easy to find their pseudoinverses

$$\begin{pmatrix} 1 & 0 \\ 0 & x \end{pmatrix}^+ = \begin{pmatrix} 1 & 0 \\ 0 & \theta(x) \end{pmatrix} = \mathrm{diag}[1, \theta(x)]$$

Here

$$\theta(x) = \begin{cases} x^{-1} & \text{for} \quad x \neq 0 \\ 0 & \text{for} \quad x = 0 \end{cases}$$

The function $\theta(x)$ will be of great importance and will be widely used in subsequent constructions of minimal pseudoinverse matrices.

Problem (1) of constructing the minimal pseudoinverse matrix $\widetilde{A}_h = \mathrm{diag}[1, \theta(\tilde{x})]$ reduces to seeking a number $\tilde{x} \in \mathbf{R}$ for which

$$1 + \theta(\tilde{x}^2) = \inf\{1 + \theta(x^2) \colon (x - H)^2 \le h^2\}$$

The graph of the function $1 + \theta(x^2)$ is depicted in Fig. 6.1. It is easily seen that $1 + \theta(\tilde{x}^2) = \min\{1 + (H - h)^{-2}, 1, 1 + (H + h)^{-2}\} = 1$, that is, $\tilde{x} = 0$.

Thus, $\widetilde{A}_h = \mathrm{diag}(1, 0) = A$. Here the equality $\|\widetilde{A}_h - A_h\| = h$ is violated. That is related to the specific configuration of the set $\mathfrak{A}_0$.

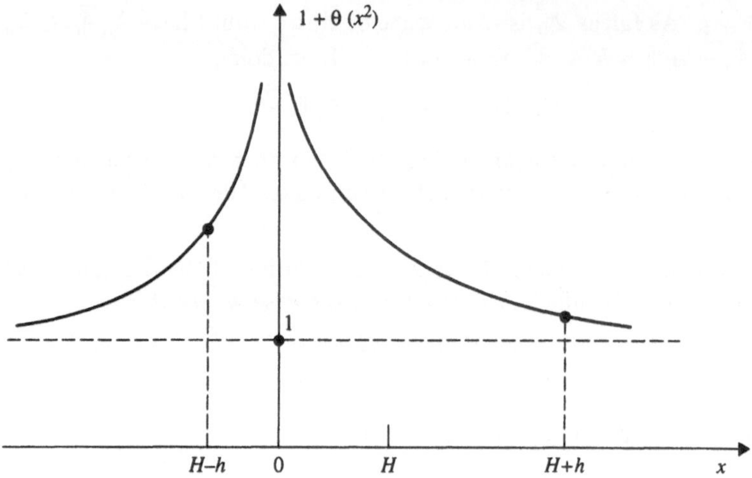

Figure 6.1 Illustration to Example 1.

Example 2 We now consider in Example 1

$$\mathfrak{A}_0 = \{\operatorname{diag}(x, y) : x, y \in \mathbf{R}\}$$

Then with every $A \in \mathfrak{A}_0$ one associates $A^+ = \operatorname{diag}[\theta(x), \theta(y)]$. Let $\|A_h\|^2 = 1 + H^2 > h^2$ (for instance, this is true for any h from the half-open interval $0 \le h < 1$). Then, by Corollary 2, since $\mathfrak{A}_0$ is a subspace of $\mathfrak{A}$, the minimal pseudoinverse matrix

$$\widetilde{A}_h^+ = \operatorname{diag}[\theta(\tilde{x}), \theta(\tilde{y})]$$

will be taken to be a solution of the extremal problem for numbers $\tilde{x}$ and $\tilde{y}$ satisfying the relation

$$\theta(\tilde{x}^2) + \theta(\tilde{y}^2) = \inf\{\theta(x^2) + \theta(y^2) : (x-1)^2 + (y-H)^2 = h^2\}$$

Such problems will be of special investigations in Section 3 (see Example 3.1). Here we restrict ourselves to writing only its solution

$$\tilde{x} = 1 + (h^2 - H^2)^{1/2} \qquad \tilde{y} = 0 \qquad 0 \le H \le h < \tfrac{1}{2}$$

Theorem 3 *If $\widetilde{A}_h$ is a minimal pseudoinverse matrix, then $\widetilde{A}_h^+$ converges to $\bar{A}^+$ as $h \to 0$.*

Proof. Observe that $\bar{A} \in \mathfrak{A}_h$ and (1) imply the estimate

$$\|\widetilde{A}_h^+\|_* \le \|\bar{A}^+\|_* \tag{5}$$

Since $A_h \to \bar{A}$ as $h \to 0$, we can deduce by Lemma 1 from the preceding estimate the assertion of the theorem. □

Theorem 3 allows one to justify the convergence of approximations $z_\eta =$

$\widetilde{A}_h^+ u_\sigma$ obtained by means of the **m.p.m.** method to a normal pseudosolution $\bar{z}$ of system (1.1) as $\eta \to 0$. Indeed, applying inequality (5) yields

$$\begin{aligned} \|z_\eta - \bar{z}\| &= \|\widetilde{A}_h^+ u_\sigma - \bar{A}^+ \bar{u}\| \le \|\widetilde{A}_h^+ u_\sigma - \widetilde{A}_h^+ \bar{u}\| + \|\widetilde{A}_h^+ \bar{u} - \bar{A}^+ \bar{u}\| \\ &\le \|\widetilde{A}_h^+\|_* \cdot \|u_\sigma - \bar{u}\| + \|\widetilde{A}_h^+ - \bar{A}^+\|_* \cdot \|\bar{u}\| \\ &\le \|\bar{A}^+\|_* \sigma + \|\widetilde{A}_h^+ - \bar{A}^+\|_* \cdot \|\bar{u}\| \end{aligned} \tag{6}$$

Taking estimate (6) to the limit as $\eta = (h, \sigma) \to 0$, we establish the convergence $z_\eta \to \bar{z}$. □

In what follows we expound certain exploratory devices for evaluating the accuracy of approximations obtained by means of the **m.p.m.** method. For a moment, we focus our attention on useful properties of the **m.p.m.** method. Some of them will be needed in the sequel.

First, it is worth noting here that the matrix $\widetilde{A}_h$ in the framework of the **m.p.m.** method is subject to the inequalities

$$\|\widetilde{A}_h - \bar{A}\| \le \|\widetilde{A}_h - A_h\| + \|A_h - \bar{A}\| \le 2h \tag{7}$$

Second, one of the important properties of the method entitled is that for all sufficiently small h its matrix and an unknown exact matrix $\bar{A}$ have one and the same rank.

Theorem 4 *Let $\widetilde{A}_h$ be an arbitrary solution of problem* (1) *for $0 \le h < h_0(\bar{A}) \equiv \|\bar{A}^+\|_*^{-1}/2$. Then*

$\mathrm{Rg}\widetilde{A}_h = \mathrm{Rg}\bar{A}$.

Before we undertake the proof of this theorem, it is reasonable to introduce some quantities which will be involved in subsequent reasonings. Set $M \equiv \min(m, n)$ and consider the singular value decompositions of the matrices $\bar{A}$ and A_h:

$$\bar{A} = \bar{U}\bar{D}\bar{V}^{\mathrm{T}} \qquad \text{and} \qquad A_h = U_h D_h V_h^{\mathrm{T}}$$

Here $\bar{U}, U_h \in \mathfrak{A}_m$ and $\bar{V}, V_h \in \mathfrak{A}_n$ stand for orthogonal matrices, while

$$\bar{D} = \mathrm{diag}(\bar{\rho}_1, \dots, \bar{\rho}_M) \in \mathfrak{A} \qquad D_h = \mathrm{diag}(\rho_1^h, \dots, \rho_M^h) \in \mathfrak{A}$$

are rectangular diagonal matrices consisting of the singular values $\bar{\rho}_k$ and ρ_k^h, $k = 1, \dots, M$, arranged in the nonincreasing order and related to the matrices $\bar{A}$ and A_h, respectively:

$$\bar{\rho}_1 \ge \cdots \ge \bar{\rho}_M \ge 0$$

and

$$\rho_1^h \ge \cdots \ge \rho_M^h \ge 0.$$

As known from Voevodin and Kuznetsov (1984), Lawson and Hanson (1974), and Section 1.6 in Chapter 1, for singular values $\rho_1 \ge \cdots \ge \rho_M$ of any matrix $A \in \mathfrak{A}$ the relations

$$\sum_{k=1}^{M} (\rho_k - \bar{\rho}_k)^2 \leq \|A - A_h\|^2 \tag{8}$$

$$\|A^+\|_*^2 = \sum_{k=1}^{M} \theta(\rho_k^2) = \sum_{k=1}^{r(A)} \rho_k^{-2} \tag{9}$$

are valid with $\rho_k = 0$ for $r(A) < k \leq M$, where $r(A) \equiv \mathrm{Rg}A$ is the rank of the matrix A.

Proof of Theorem 4. Let $\tilde{\rho}_1, \dots, \tilde{\rho}_M$ exhaust all singular values of a matrix $\widetilde{A}_h$ obtained by means of the **m.p.m.** method. Accepting $t \equiv \mathrm{Rg}\widetilde{A}_h$, $r \equiv \mathrm{Rg}\bar{A}$ and $t < r$, we deduce from inequalities (7)–(8) that

$$\bar{\rho}_r^2 \leq \sum_{k=1}^{t} (\tilde{\rho}_k - \bar{\rho}_k)^2 + \sum_{k=t+1}^{r} \bar{\rho}_k^2 = \sum_{k=1}^{M} (\tilde{\rho}_k - \bar{\rho}_k)^2 \leq \|\widetilde{A}_h - \bar{A}\|^2 \leq 4h^2$$

and relation (9), in turn, yields

$$\bar{\rho}_r^{-2} \leq \|\bar{A}^+\|_*^2$$

From both estimates for $\bar{\rho}_r$ it follows that $h^2 \geq \|\bar{A}^+\|^{-2}/4$, which contradicts the hypothesis of the theorem. Therefore, the inequality $t \geq r$ should be valid. When the strict inequality $t > r$ is attained, we derive from (7)–(9) with the aid of estimate (5) the chain of relations

$$\tilde{\rho}_{r+1}^2 = (\tilde{\rho}_{r+1} - \bar{\rho}_{r+1})^2 \leq \sum_{k=1}^{M} (\tilde{\rho}_k - \bar{\rho}_k)^2 \leq 4h^2$$

$$\tilde{\rho}_{r+1}^{-2} \leq \sum_{k=1}^{t} \tilde{\rho}_k^{-2} = \|\widetilde{A}_h^+\|_*^2 \leq \|\bar{A}^+\|_*^2$$

Combining this result with the preceding, we would have $h^2 \geq h_0^2(\bar{A})$, which disagrees with the above restrictions on h. This provides support for the view that $t = r$. □

Corollary 3 *The singular values $\tilde{\rho}_k$ of the matrix $\widetilde{A}_h$ vanish for $k > r$ and do not vanish for $1 \leq k \leq r$ provided that $0 \leq h < h_0(\bar{A})$.*

Corollary 4 *By exactly the same reasoning as in the proof of Theorem 4 for the case $t < r$,*

$$RgA_h \geq Rg\bar{A}$$

By analogy with the proof of Theorem 4 one can derive the following corollary.

Corollary 5 *If $\|\check{A}_h - \bar{A}\| \leq qh$, $q = const \geq 1$, and $\|\check{A}_h^+\|_* \leq \|\bar{A}^+\|_*$ for a one-parameter family of matrices $\{\check{A}_h, h \geq 0\}$, then*

$$Rg\check{A}_h = Rg\bar{A}$$

provided that $0 \leq h < \|\bar{A}^+\|_*^{-1}/q$.

Evidently, Theorem 4 yields $\mathrm{Rg}\widetilde{A}_h^+ = \mathrm{Rg}\bar{A}^+$ for $0 \le h < h_0(\bar{A})$ as well. In view of this, the **m.p.m.** method is distinguished from many existing algorithms for the stable solution of SLAEs in which the approximation of pseudoinverses is essentially used. For instance, the method of Vaĭnikko and Veretennikov (1986) and of Gilyazov and Morozov (1984) is based on the following approximation to $\bar{A}^+$:

$$B_h = (hI + A_h^{\mathrm{T}} A_h)^{-1} A_h^{\mathrm{T}} A_h A_h^{\mathrm{T}} (hE + A_h A_h^{\mathrm{T}})^{-1}$$

Obviously, the matrix B_h may have the rank other than $\mathrm{Rg}\bar{A}$ even if h may be made arbitrarily small. To make sure of it, the matrices A_h and $\bar{A}$ from Example 1 suit perfectly.

Bearing in mind the conclusions of Theorem 4, one can estimate the rate of convergence $\widetilde{A}_h^+ \to \bar{A}^+$ as $h \to 0$ that has been established in Theorem 3. Much progress has been achieved by results of Wedin (1973).

Lemma 2 *Let $A, B \in \mathfrak{A}$ and $RgA = RgB$. Then*

$$\|A^+ - B^+\|_* \le L\|A - B\| \cdot \|A^+\|_* \cdot \|B^+\|_* \tag{10}$$

if a constant L is defined as follows: $L = 1$ when $RgA = M$ and $L = \sqrt{2}$ when $RgA < M$.

Since Theorem 4 guarantees $\mathrm{Rg}\widetilde{A}_h = \mathrm{Rg}\bar{A}$ as long as $0 \le h < h_0(\bar{A})$, combination of Lemma 2 and relations (5) and (7) gives the following estimate for covergence rate.

Theorem 5 *For $0 \le h < h_0(\bar{A})$*

$$\|\widetilde{A}_h^+ - \bar{A}^+\|_* \le \sqrt{2}\|\widetilde{A}_h - \bar{A}\| \cdot \|\widetilde{A}_h^+\|_* \cdot \|\bar{A}^+\|_* \le 2\sqrt{2}h\|\bar{A}^+\|_*^2$$

By considering the simplest examples it is easy to make sure that the last estimate is the best possible in order of its ingredients h and $\|\bar{A}^+\|_*$ whereas the constant $2\sqrt{2}$ can be improved as $h \to 0$. The value of this constant will be of significance in establishing the accuracy properties of the **m.p.m** method. The following lemma allows us to refine estimate (10).

Lemma 3 *Let $A, B \in \mathfrak{A}$, $RgA = RgB$ and $\|A - B\| \cdot \|A^+\|_* < 1$. Then*

$$\|A^+ - B^+\|_* \le \frac{\|A - B\| \cdot \|A^+\|_*^2}{(1 - \|A - B\| \cdot \|A^+\|_*)^3}$$

Proof. Let a matrix A admit the singular value decomposition $A = UPV^{\mathrm{T}}$, where

$$P = \mathrm{diag}(p_1, \dots, p_t, 0, \dots, 0) \in \mathfrak{A}$$

and

$$t \equiv \mathrm{Rg}A.$$

Owing to the orthogonality of the matrices U and V and the equality

$A^+ = VP^+U^{\mathrm{T}}$ we arrive at the set of relations:

$$\begin{aligned}
\|A - B\| &= \|UPV^{\mathrm{T}} - B\| = \|P - U^{\mathrm{T}}BV\| = \|P - Q\| \\
& \quad Q \equiv U^{\mathrm{T}}BV \\
\|A^+ - B^+\|_* &= \|VP^+U^{\mathrm{T}} - B^+\|_* = \|P^+ - V^{\mathrm{T}}B^+U\|_* \\
&= \|P^+ - (U^{\mathrm{T}}BV)^+\|_* = \|P^+ - Q^+\|_* \\
& \quad \|A^+\|_* = \|P^+\|_*
\end{aligned} \tag{11}$$

Since $\mathrm{Rg}P = t = \mathrm{Rg}B = \mathrm{Rg}Q$, the matrices P and Q can be represented in the block forms

$$P = [W \vdots 0] \qquad W \in \mathfrak{A}_{m\times t} \qquad 0 \in \mathfrak{A}_{m\times(n-t)} \qquad \mathrm{Rg}W = t$$

$$Q = [R \vdots S] \qquad R \in \mathfrak{A}_{m\times t} \qquad S \in \mathfrak{A}_{m\times(n-t)} \qquad \mathrm{Rg}R = \mathrm{Rg}Q = t$$

Due to the linear dependence of columns of the matrix S upon those of the matrix R (the theorem on basis minors), the equality $S = RR^+S$ holds true. As shown in Albert (1972), under the above conditions we might have

$$P^+ = \begin{pmatrix} W^+ \\ 0 \end{pmatrix} \qquad Q^+ = \begin{pmatrix} R^+ - R^+SJ \\ J \end{pmatrix}$$

where

$$\begin{aligned}
J &\equiv KZ^{\mathrm{T}}R^+ \in \mathfrak{A}_{(n-t)\times m} \qquad & K &\equiv (E + Z^{\mathrm{T}}Z)^{-1} \\
Z &\equiv R^+S \in \mathfrak{A}_{n\times(n-t)} \qquad & E &\equiv \mathrm{diag}(1, \ldots, 1) \in \mathfrak{A}_{(n-t)\times(n-t)}
\end{aligned}$$

It follows from the foregoing that

$$\begin{aligned}
\|P^+ - Q^+\|_*^2 &= \|W^+ - R^+ + R^+SJ\|^2 + \|J\|^2 \\
&\le (\|W^+ - R^+\| + \|R^+SJ\|)^2 + \|J\|^2
\end{aligned}$$

Let us evaluate the right-hand side of the last inequality. For this, one way of proceeding is to derive the chain of relations

$$\begin{aligned}
\|J\|^2 &= \|KZ^{\mathrm{T}}R^+\|^2 \le \|(E + Z^{\mathrm{T}}Z)^{-1}Z^{\mathrm{T}}\|^2 \cdot \|R^+\|^2 \\
&= \|R^+\|^2 \sum_{i=1}^{r(Z)} \frac{\mu_i^2}{(1 + \mu_i^2)^2} \\
&\le \|R^+\|^2 \sum_{i=1}^{r(Z)} \mu_i^2 = \|R^+\|^2 \cdot \|Z\|^2 \\
&= \|R^+\|^2 \cdot \|R^+S\|^2 \le \|R^+\|^4 \cdot \|S\|^2
\end{aligned}$$

Here μ_i $(i = 1, \ldots, r(Z))$ indicate nonzero singular values of the matrix Z and $r(Z) \equiv \mathrm{Rg}Z$.

The collection of the above estimates yields

$$\|R^+SJ\| \le \|R^+\| \cdot \|S\| \cdot \|J\| \le \|R^+\|^3 \cdot \|S\|^2$$

and, therefore,

$$\begin{aligned} \|P^+ - Q^+\|_*^2 \; &\leq \; (\|W^+ - R^+\| \\ &\quad + \|R^+\|^3 \cdot \|S\|^2)^2 \\ &\quad + \|R^+\|^4 \cdot \|S\|^2 \end{aligned} \tag{12}$$

To majorize $\|W^+ - R^+\|$ and $\|R^+\|$ we appeal to Lemma 2 providing for W and R, as for matrices of the total rank t,

$$\|W^+ - R^+\| \leq \|W^+\| \cdot \|R^+\| \cdot \|W - R\| \tag{13}$$

The condition of the lemma and (11) imply

$$\begin{aligned} \|W - R\|^2 \cdot \|W^+\|^2 \; &\leq \; (\|W - R\|^2 + \|S\|^2)\|W^+\|^2 \\ &= \; \|P - Q\|^2 \cdot \|P^+\|_*^2 \\ &= \; \|A - B\|^2 \cdot \|A^+\|_*^2 < 1 \end{aligned}$$

With these relations established, we derive from (13) the estimate

$$\|R^+\| \leq \frac{\|W^+\|}{1 - \|W - R\| \cdot \|W^+\|} \tag{14}$$

For the sake of brevity we keep the following notations

$$\begin{aligned} &\|W - R\| \equiv \varepsilon \qquad\qquad \|S\| \equiv \omega \\ &\|A - B\|^2 = \|P - Q\|^2 = \varepsilon^2 + \omega^2 \equiv H^2 \end{aligned}$$

With the aid of (11)–(12) along with (13)–(14) we establish the chain of inequalities

$$\begin{aligned} \|A^+ - B^+\|_*^2 \; &= \; \|P^+ - Q^+\|_*^2 \leq [\|W^+\|^2 \varepsilon (1 - \varepsilon\|W^+\|)^{-1} \\ &\quad + \|W^+\|^3 \omega^2 (1 - \varepsilon\|W^+\|)^{-3}]^2 \\ &\quad + \|W^+\|^4 \omega^2 (1 - \varepsilon\|W^+\|)^{-4} \\ &= \; \frac{\|W^+\|^4}{(1 - \varepsilon\|W^+\|)^4} \Big[\varepsilon^2 (1 - \varepsilon\|W^+\|)^2 + \omega^2 \\ &\quad + 2\varepsilon\omega^2 \|W^+\| + \frac{\omega^4 \|W^+\|^2}{(1 - \varepsilon\|W^+\|)^2} \Big] \\ &\leq \; \frac{\|W^+\|^4 (\varepsilon^2 + \omega^2)}{(1 - \varepsilon\|W^+\|)^4} \Big[1 + \frac{\varepsilon^4 \|W^+\|^2}{\varepsilon^2 + \omega^2} + \frac{2\varepsilon\omega^2}{\varepsilon^2 + \omega^2} \|W^+\| \\ &\quad + \frac{\omega^4}{\varepsilon^2 + \omega^2} \frac{\|W^+\|^2}{(1 - \varepsilon\|W^+\|)^2} \Big] \end{aligned}$$

$$\leq \frac{\|W^+\|^4 H^2}{(1-H\|W^+\|)^4}\Big[1+2\varepsilon\|W^+\|$$
$$+\frac{\varepsilon^4+\omega^4}{\varepsilon^2+\omega^2}\frac{\|W^+\|^2}{(1-H\|W^+\|)^2}\Big]$$
$$\leq \frac{H^2\|W^+\|^4}{(1-H\|W^+\|)^4}\left[1+\frac{2H\|W^+\|}{1-H\|W^+\|}+\frac{H^2\|W^+\|^2}{(1-H\|W^+\|)^2}\right]$$
$$= \frac{H^2\|W^+\|^4}{(1-H\|W^+\|)^4}\left[1+\frac{H\|W^+\|}{1-H\|W^+\|}\right]^2$$
$$= \frac{H^2\|W^+\|^4}{(1-H\|W^+\|)^6}=\frac{H^2\|A^+\|_*^4}{(1-H\|A^+\|_*)^6}$$

agreeing with trivial inequalities

$$1\leq(1-\varepsilon\|W^+\|)^{-1}\leq(1-H\|W^+\|)^{-1} \qquad \varepsilon^2/(\varepsilon^2+\omega^2)\leq 1$$

Thus, the lemma is completely proved. □

Lemma 3 and Theorem 4 imply one important result.

Theorem 6 *For* $0\leq h<h_0(\bar{A})$ *the following estimate is valid:*

$$\|\widetilde{A}_h^+-\bar{A}^+\|_*\leq\frac{2h\|\bar{A}^+\|_*^2}{(1-2h\|\bar{A}^+\|_*)^3} \tag{15}$$

It is clear that the main term of estimate (15) is of the form

$$\|\widetilde{A}_h^+-\bar{A}^+\|_*\asymp 2h\|\bar{A}^+\|_*^2$$

as $h\to 0$. The constant 2 cannot be asymptotically improved. This means that there are matrices $\bar{A}$ and A_h with $\|\bar{A}-A_h\|=h$ such that

$$\lim_{h\to+0}\frac{\|\widetilde{A}_h^+-\bar{A}^+\|_*}{h\|\bar{A}^+\|_*^2}=2$$

The following example may help to verify the latter.

Example 3 Let $\bar{A}=\mathrm{diag}(1,0)\in\mathfrak{A}_2$ and the set $\mathfrak{A}_0$ be expressed by

$$\mathfrak{A}_0=\left\{\begin{pmatrix}1 & x\\ x & x^2\end{pmatrix}: x\in\mathbf{R}\right\}$$

Because of its form, A_h will be taken to be

$$A_h=\begin{pmatrix}1 & H\\ H & H^2\end{pmatrix} \qquad 2H^2+H^4\equiv h^2,\quad 0\leq h<1\quad H\geq 0$$

Evidently, $\|\bar{A}-A_h\|^2=h^2$. For any matrix from the set $\mathfrak{A}_0$ its pseudoinverse looks as follows:

$$A^+=\begin{pmatrix}1 & x\\ x & x^2\end{pmatrix}^+=\lim_{\alpha\to+0}(\alpha E+A^{\mathrm{T}}A)^{-1}A^{\mathrm{T}}=\frac{1}{(1+x^2)^2}\begin{pmatrix}1 & x\\ x & x^2\end{pmatrix}$$

where

$$\|A^+\|_*^2 = \frac{1+2x^2+x^4}{(1+x^2)^4} = \frac{1}{(1+x^2)^2}$$

The set $\mathfrak{A}_h$ is specified by means of the inequality

$$\|A - A_h\|^2 = 2(x-H)^2 + (x^2 - H^2)^2 \leq h^2 = 2H^2 + H^4$$

From here it is easily seen that $A \in \mathfrak{A}_h$ if $0 \leq x \leq x(H)$, where $x(H)$ is a unique real root to the equation $x^3 + 2x(1-H^2) - 4H = 0$. Notice that $x(H) \asymp 2H$ as $H \to 0$. Then problem (1) of constructing the matrix of the **m.p.m.** method amounts to finding a number $\tilde{x}$ for which

$$(1+\tilde{x}^2)^{-2} = \inf\left\{(1+x^2)^{-2} \colon 0 \leq x \leq x(H)\right\}$$

Therefore, $\tilde{x} = x(H)$ and, in view of the equality $\|\bar{A}^+\|_* = 1$,

$$\begin{aligned} \|\widetilde{A}_h^+ - \bar{A}^+\|_*^2 &= \left(1 - \frac{1}{(1+\tilde{x}^2)^2}\right)^2 \\ &\quad + \frac{2\tilde{x}^2}{(1+\tilde{x}^2)^4} + \frac{\tilde{x}^4}{(1+\tilde{x}^2)^4} \\ &= \frac{2\tilde{x}^2 + \tilde{x}^4}{(1+\tilde{x}^2)^2}\|\bar{A}^+\|_*^4 \end{aligned}$$

yielding

$$\lim_{h\to+0} \frac{\|\widetilde{A}_h^+ - \bar{A}^+\|_*}{h\|\bar{A}^+\|_*^2} = \lim_{H\to 0}\sqrt{\frac{2x^2(H) + x^4(H)}{(2H^2+H^4)[1+x^2(H)]^2}} = 2$$

On the strength of Theorem 6, we proceed to estimate with the aid of inequality (6) the accuracy of approximations z_η obtained by means of the **m.p.m.** method to a normal pseudosolution $\bar{z}$.

Theorem 7 *The estimate is true:*

$$\begin{aligned} \|z_\eta - \bar{z}\| &\leq \|\bar{A}^+\|_*\sigma + \|\widetilde{A}_h^+ - \bar{A}^+\|_* \cdot \|\bar{u}\| \\ &\leq \|\bar{A}^+\|_*\left[\sigma + \frac{2h\|\bar{A}^+\|_*\|\bar{u}\|}{(1-2h\|\bar{A}^+\|_*)^3}\right] \end{aligned} \tag{16}$$

From estimate (16) we draw the conclusion that the approximation z_η is of optimal accuracy order $O(\sigma + h)$ disregarding to the fact whether system (1.1) is compatible or not. Moreover, estimate (16) allows one to compare the accuracy of the **m.p.m.** method with that of another method from the work of Vaĭnikko and Veretennikov (1986) or from the work of Gilyazov and Morozov (1984). The last estimate is presented in formula (1.9).

In conclusion we emphasize that problem (1) of the **m.p.m.** method falls within the category of possible versions of the generalized discrepancy method from Section 3.5 in Chapter 3 with $\Omega(A) = \|A^+\|_*$.

However, functionals like $\Omega(A)$ have no properties of semicontinuity and compactness on the space $\mathfrak{A}$ in contrast to those of Section 2.2 in Chapter 2. So, the theory of the generalized discrepancy method outlined in Chapter 3 does not apply directly to the **m.p.m.** method.

But such a functional $\Omega(A)$ possesses on the space $\mathfrak{A}$ some specific properties such as 'bounded continuity' and 'bounded compactness' justified in Lemma 2 and Corollary 1. These properties ensure convergence of the **m.p.m.** method (see Theorem 3).

Remark 1 The reason why it is not always necessary to be specific about which norm is being used is that many facts expressed in terms of one norm imply the corresponding facts in which a different norm is used. The **m.p.m.** method admits some modifications in which alternative norms of the finite-dimensional spaces $\mathfrak{A}$ and $\mathfrak{A}^*$ are adopted in the general setting instead of Euclidean norms $\|\cdot\|$ and $\|\cdot\|_*$. The new norms of $\mathfrak{A}$ and $\mathfrak{A}^*$ must be subordinate to the norms of the spaces $\mathbf{R}^n$ and $\mathbf{R}^m$. The assertions of Lemma 1 and Theorems 1–3 remain valid for such modifications of the **m.p.m.** method since all the norms on those spaces $\mathfrak{A}$ and $\mathfrak{A}^*$ are equivalent. Theorems 4–7 also have the appropriate analogues. However, the constraints on h in their conditions and the constants in the estimates are somewhat different.

6.3 Constructing the matrix of the m.p.m. method

The solution of problem (2.1) is a keystone in realizing the **m.p.m.** method. For the same reason as above, it suffices to find any solution $\widetilde{A}_h$ of problem (2.1) that serves as a necessary background for further constructions of the approximate solution $z_\eta = \widetilde{A}_h^+ u_\sigma$ of system (1.1). In this section, we attempt a solution of problem (2.1) in a special form. This approach is best suited for the case where $\mathfrak{A}_0 = \mathfrak{A}$. In other words, the *a priori* set $\mathfrak{A}_0$ will be identical with the entire space $\mathfrak{A}$.

It is interesting to study an auxiliary problem of mathematical programming in which for a fixed h and the singular values $\rho_k^h (k = 1, \dots, M)$ of a given matrix A_h it is required to find numbers $\hat{\rho}_1, \dots, \hat{\rho}_M$ satisfying the monotonicity condition

$$\hat{\rho}_1 \geq \cdots \geq \hat{\rho}_M \geq 0$$

and

$$\sum_{k=1}^{M} \theta(\hat{\rho}_k^2) = \inf\left\{ \sum_{k=1}^{M} \theta(\rho_k^2) \colon \rho_1 \geq \rho_2 \geq \cdots \geq \rho_M \geq 0, \right.$$
$$\left. \sum_{k=1}^{M} (\rho_k - \rho_k^h)^2 = h^2 \right\} \tag{1}$$

Along with this problem, we look for numbers $\hat{\rho}_1, \dots, \hat{\rho}_M$ such that $\hat{\rho}_1 \geq$

$\cdots \geq \hat{\rho}_M \geq 0$ and

$$\sum_{k=1}^{M} \theta(\hat{\rho}_k^2) = \inf\left\{ \sum_{k=1}^{M} \theta(\rho_k^2) \colon \rho_1 \geq \rho_2 \geq \cdots \geq \rho_M \geq 0, \right.$$
$$\left. \sum_{k=1}^{M} (\rho_k - \rho_k^h)^2 \leq h^2 \right\} \tag{2}$$

In the next lemma, we establish an interconnection between problems (1) and (2).

Lemma 1 *Problems* (1) *and* (2) *are solvable and equivalent under the condition* $\|A_h\| > h$.

Proof. Holding the data (A_h, h) fixed, we refer to the matrices U_h and V_h arising in the singular value decomposition of the matrix A_h by means of which we can specify the set of matrices

$$\mathfrak{A}(A_h) = \{A \colon A \in \mathfrak{A}, A = U_h D V_h^{\mathrm{T}}, D = \operatorname{diag}(\rho_1, \ldots, \rho_M) \in \mathfrak{A},$$
$$\rho_1 \geq \cdots \geq \rho_M \geq 0\}$$

With the aid of obvious relations

$$\|A - A_h\|^2 = \|U_h(D - D_h)V_h^{\mathrm{T}}\|^2 = \|D - D_h\|^2 = \sum_{k=1}^{M} (\rho_k - \rho_k^h)^2$$

which are valid for any matrix from the set $\mathfrak{A}(A_h)$, and equality (2.9) problems (1) and (2) can be written in the equivalent form:

$$\inf\{\|A^+\|_*^2 \colon A \in \mathfrak{A}(A_h), \|A - A_h\|^2 = h^2\} \tag{1$'$}$$
$$\inf\{\|A^+\|_*^2 \colon A \in \mathfrak{A}(A_h), \|A - A_h\|^2 \leq h^2\} \tag{2$'$}$$

Such statements of problems (1′) and (2′) are similar in form to problems (2.4) and (2.1), respectively, if we agree to consider h to be fixed and $\mathfrak{A}_0 = \mathfrak{A}(A_h)$.

Therefore, the existence of solutions of problems (1′) and (2′) and their equivalence follow immediately from Theorems 2.1 and 2.2 provided the condition

$$\|A_h\|^2 = \sum_{k=1}^{M} (\rho_k^h)^2 > h^2 \tag{3}$$

holds. □

Theorem 1 *Let under condition* (3) *the numbers* $\hat{\rho}_1, \ldots, \hat{\rho}_M$ *constitute a solution of problem* (1). *Then the matrix* $\widehat{A}_h \equiv U_h \widehat{D}_h V_h^{\mathrm{T}}$, *where*

$\widehat{D}_h \equiv \operatorname{diag}(\hat{\rho}_1, \ldots, \hat{\rho}_M) \in \mathfrak{A}$, gives a solution of the extremal problem (2.1) of the **m.p.m.** method with $\mathfrak{A}_0 = \mathfrak{A}$.

So, Theorem 1 provides the framework for constructing one of the matrices of the **m.p.m.** method in terms of the matrices U_h and V_h from the singular value decomposition of A_h and the diagonal matrix $\widehat{D}_h$ the elements of which constitute a solution of problem (1).

Proof. We take into consideration any matrix $A \in \mathfrak{A}$ and its singular value decomposition

$$A = UDV^{\mathrm{T}} \qquad D = \mathrm{diag}(\rho_1, \dots, \rho_M)$$

By formula (2.8),

$$\|D - D_h\|^2 = \sum_{k=1}^{M} (\rho_k - \rho_k^h)^2 \leq \|A - A_h\|^2 \qquad \forall A \in \mathfrak{A} \tag{4}$$

As a preliminary, we introduce the two sets

$$\mathfrak{A}_h = \{A \in \mathfrak{A}\colon \|A - A_h\| \leq h\}$$

and $\mathfrak{B}_h$ consisting of matrices $A = UDV^{\mathrm{T}}$ with $\|D - D_h\| \leq h$ and having the form

$$\mathfrak{B}_h = \{A \in \mathfrak{A}\colon A = UDV^{\mathrm{T}}, \|D - D_h\| \leq h\}$$

According to (4) the inequality $\|A - A_h\| \leq h$ implies $\|D - D_h\| \leq h$. This means that $\mathfrak{A}_h \subset \mathfrak{B}_h$ and confirms the validity of the relations

$$\begin{aligned}
\inf\{\|A^+\|_*^2\colon A \in \mathfrak{A}_h\} &\geq \inf\{\|A^+\|_*^2\colon A \in \mathfrak{B}_h\} \\
&= \inf\{\|D^+\|_*^2\colon \|D - D_h\| \leq h\} \\
&= \inf\Big\{\sum_{k=1}^{M} \theta(\rho_k^2)\colon \rho_1 \geq \dots \geq \rho_M \geq 0, \\
&\qquad \sum_{k=1}^{M} (\rho_k - \rho_k^h)^2 \leq h^2\Big\}
\end{aligned} \tag{5}$$

where one possible variant of equality (2.9), namely

$$\|A^+\|_*^2 = \|D^+\|_*^2 = \sum_{k=1}^{M} \theta(\rho_k^2)$$

and relation (4) are taken into account. By Lemma 1, the last greatest lower bound in (5) is attained for the numbers $\hat{\rho}_1, \dots, \hat{\rho}_M$, being a solution of problem (1), so that (5) implies the inequality

$$\begin{aligned}
\|A^+\|_*^2 &\geq \inf\{\|A^+\|_*^2\colon A \in \mathfrak{A}_h\} \\
&\geq \sum_{k=1}^{M} \theta(\hat{\rho}_k^2) = \|\widehat{D}_h^+\|_*^2 \qquad \forall A \in \mathfrak{A}_h
\end{aligned} \tag{6}$$

Now, we introduce the matrix $\widehat{A}_h \equiv U_h \widehat{D}_h V_h^{\mathrm{T}}$, which satisfies due to (1) the chain of relations

$$\begin{aligned}
\|\widehat{A}_h - A_h\|^2 &= \|U_h \widehat{D}_h V_h^{\mathrm{T}} - U_h D_h V_h^{\mathrm{T}}\|^2 \\
&= \|\widehat{D}_h - D_h\|^2 = \sum_{k=1}^{M} (\hat{\rho}_k - \rho_k^h)^2 = h^2
\end{aligned}$$

which implies $\widehat{A}_h \in \mathfrak{A}_h$. Then, according to (6),

$$\|\widehat{A}_h^+\|_*^2 = \|\widehat{D}_h^+\|_*^2 \leq \|A^+\|_*^2 \qquad \forall A \in \mathfrak{A}_h$$

This provides support for the view that the matrix $\widehat{A}_h$ is just a solution of problem (2.1). □

From Theorem 1 we now know what can be done to produce the matrix of the **m.p.m.** method, whose implementation necessitates imposing the singular value decomposition of a given matrix A_h by means of which we seek a solution $(\hat{\rho}_1, \dots, \hat{\rho}_M)$ of problem (1) and construct the matrix $\widehat{A}_h$ of the **m.p.m.** method. Algorithmic features concerning practical realizations of the singular value decomposition will be discussed later. Our first step is to find a solution of problem (1) in the explicit form. There seem to be at least two possible approaches to the solution of this problem.

The first tack is concerned with the use of the Lagrange function for problem (1). The Lagrange function is free to be chosen in any convenient way:

$$\begin{aligned} M^\lambda[\rho_1, \dots, \rho_M] &\equiv \lambda \sum_{k=1}^{M} \theta(\rho_k^2) + \sum_{k=1}^{M} (\rho_k - \rho_k^h)^2 \\ &\lambda \geq 0 \quad \rho_1, \dots, \rho_M \geq 0 \end{aligned} \tag{7}$$

For the cases of interest we may set the extremal problem related to numbers $\rho_1(\lambda), \dots, \rho_M(\lambda) \geq 0$ such that

$$M^\lambda[\rho_1(\lambda), \dots, \rho_M(\lambda)] = \inf\{M^\lambda[\rho_1, \dots, \rho_M] \colon \rho_1, \dots, \rho_M \geq 0\} \tag{8}$$

if $\lambda \geq 0$ is held fixed.

Lemma 2 *Problem* (8) *is solvable for any* $\lambda \geq 0$. *Each of its solutions can be determined by the relations*

$$\rho_k(\lambda) = \begin{cases} \rho_k^h x_k(\lambda) & \text{if } \ 0 < \lambda \leq \lambda_k \\ 0 & \text{if } \ \ \lambda \geq \lambda_k \end{cases} \tag{9}$$

$$\rho_k(0) = \rho_k^h \qquad k = 1, \dots, M \tag{9'}$$

where $\lambda_k \equiv 27(\rho_k^h)^4/16$ *and* $x_k(\lambda)$ *is the root to the equation*

$$x^4 - x^3 = \lambda(\rho_k^h)^{-4} \tag{10}$$

on the segment $[1, \frac{3}{2}]$. *The values* $\rho_k(\lambda)$ *given by* (9) *and* (9′) *possess the monotonicity property* $\rho_1(\lambda) \geq \dots \geq \rho_M(\lambda) \geq 0$.

To prove Lemma 2 and some subsequent lemmas, we first establish certain properties of a positive solution to the one-parameter equation

$$x^4 - x^3 = \lambda\rho^{-4} \tag{10'}$$

where $\rho > 0$ is a fixed number and $\lambda \geq 0$ is a parameter. At first glance, the trivial examination shows that equation (10′) possesses a unique positive solution $x(\lambda)$ for each $\lambda \geq 0$ and $x(\lambda) \geq 1$. By means of standard methods of the theory of functions, it is not difficult to expose some properties of the function $x(\lambda)$. A brief survey of these properties is presented below.

(1) The function $x(\lambda)$ is continuous and infinite-differentiable for $\lambda \geq 0$. In particular,

$$x'(\lambda) = \frac{1}{\rho^4(4x^3 - 3x^2)} \qquad x''(\lambda) = -\frac{6(2x-1)}{\rho^8 x^5 (4x-3)^3}$$

(2) The function $x(\lambda)$ is monotonically increasing for $\lambda \geq 0$ with the values $x(0) = 1$ and $x(+\infty) = +\infty$.

(3) The function $x(\lambda)$ is concave for $\lambda \geq 0$.

(4) The variable $x \geq 1$ solving the preceding equation for $\lambda \geq 0$ and $\rho > 0$ depends continuously on the totality of its arguments (λ, ρ).

Proof of Lemma 2. Because of the additive form of function (7) our testing reduces to the minimum analysis for the function of one variable

$$\mathcal{F}(\rho) \equiv \lambda\theta(\rho^2) + (\rho - \rho_k^h)^2 \qquad \rho \geq 0 \quad k = 1, \dots, M$$

which is discontinuous at the point $\rho = 0$. Before taking up the general case, we suppose that $\rho_k^h > 0$ for any fixed k under consideration. By merely setting $x = \rho/\rho_k^h$ we deduce that

$$\mathcal{F}(\rho) = \begin{cases} (\rho_k^h)^2 & \text{if } \rho = 0 \\ \varphi(x) & \text{if } \rho > 0 \end{cases}$$

with

$$\varphi(x) \equiv \lambda(\rho_k^h)^{-2} x^{-2} + (\rho_k^h)^2 (x-1)^2 \qquad x > 0$$

The reader is invited to check that the function $\varphi(x)$ has a unique point $x_k(\lambda)$ of global minimum. As a result of recovering from equation (10), it does follow the inequality $x_k(\lambda) \geq 1$. By the same token,

$$\begin{aligned} \varphi_{\min} &\equiv \inf\{\varphi(x) \colon x > 0\} = \varphi[x_k(\lambda)] \\ &= \lambda(\rho_k^h)^{-2} x_k^{-2}(\lambda) + (\rho_k^h)^2 [x_k(\lambda) - 1]^2 \\ &= (\rho_k^h)^{-2}\{\lambda(\rho_k^h)^{-4} x_k^{-2}(\lambda) + [x_k(\lambda) - 1]^2\} \\ &= (\rho_k^h)^2\{x_k^2(\lambda) - x_k(\lambda) + [x_k(\lambda) - 1]^2\} \\ &= (\rho_k^h)^2[2x_k^2(\lambda) - 3x_k(\lambda) + 1] \end{aligned}$$

This provides support for further manipulations

$$\begin{aligned} \mathcal{F}_{\min} &\equiv \inf\{\mathcal{F}(\rho) \colon \rho \geq 0\} = \min\{\mathcal{F}(0), \varphi[x_k(\lambda)]\} \\ &= \min\{\mathcal{F}(0), \mathcal{F}[\rho_k^h x_k(\lambda)]\} \\ &= (\rho_k^h)^2 \min\{1, 2x_k^2(\lambda) - 3x_k(\lambda) + 1\} \end{aligned} \tag{11}$$

m.p.m. method. Formula (9) with $\lambda = \lambda(h)$ implies that $\hat{\rho}_k = \rho_k[\lambda(h)] = 0$ for $t(h)+1 \le k \le M$ and $\hat{\rho}_k > 0$ for $1 \le k \le t(h)$. On the other hand, by Theorem 2.4, the conditions $0 < h < \|\bar{A}^+\|_*^{-1}/3$ guarantee the validity of the equality $t(h) = \mathrm{Rg}\widehat{A}_h = \mathrm{Rg}\bar{A}$ we must prove.

In the second case, the equality $\lambda(h) = \lambda_{t+1}$ holds true for all h which interest us. Then the number $\lambda(h)$ being a discontinuity point of function (12) ensures

$$\beta[\lambda(h)-0] \le h^2 \le \beta[\lambda(h)+0] \tag{17}$$

and, in view of (13),

$$\begin{aligned}\beta[\lambda(h)-0] &= \beta[\lambda(h)+0] - \tfrac{3}{4}q_{t+1}(\rho_{t+1}^h)^2 \\ &= \sum_{k=1}^{t}(\rho_k^h)^2\left[x_k[\lambda(h)]-1\right]^2 \\ &\quad + \tfrac{1}{4}q_{t+1}(\rho_{t+1}^h)^2 + \sum_{k=t+2}^{M}(\rho_k^h)^2 \end{aligned} \tag{18}$$

The number q_{t+1} indicates the multiplicity of the singular value $\rho_{t(h)+1}^h$ of the matrix A_h. Relations (17) and (18) together imply

$$\tfrac{1}{4}q_{t+1}(\rho_{t+1}^h)^2 \le \beta[\lambda(h)-0] \le h^2$$

which, in turn, provide in accordance with (18) the following manipulations

$$\begin{aligned}\beta[\lambda(h)+0] &= \sum_{k=1}^{M}\left[\rho_k[\lambda(h)+0]-\rho_k^h\right]^2 \\ &= \beta[\lambda(h)-0] + \tfrac{3}{4}q_{t+1}(\rho_{t+1}^h)^2 \\ &\le h^2 + 3h^2 = 4h^2 \end{aligned} \tag{19}$$

Here the numbers $\rho_k[\lambda(h)+0] \equiv \rho_k^*$ can be found from (9).

Furthermore, combination of Minkowskiĭ's inequality and (19) gives

$$\begin{aligned}\left\{\sum_{k=1}^{M}(\rho_k^* - \bar{\rho}_k)^2\right\}^{1/2} &\le \left\{\sum_{k=1}^{M}(\rho_k^* - \rho_k^h)^2\right\}^{1/2} + \left\{\sum_{k=1}^{M}(\rho_k^h - \bar{\rho}_k)^2\right\}^{1/2} \\ &\le \beta^{1/2}[\lambda(h)+0] + h \le 3h \end{aligned} \tag{20}$$

By Lemma 2 the numbers ρ_k^* $(k = 1, \dots, M)$ constitute what is called a solution of problem (8) for $\lambda = \lambda(h)$. Therefore, putting inequality (17) together with (2.8)–(2.9) we arrive at the chain of relations

$$\begin{aligned}
\lambda(h) \sum_{k=1}^{M} \theta[(\rho_k^*)^2] + h^2 &\le \lambda(h) \sum_{k=1}^{M} \theta[(\rho_k^*)^2] + \beta[\lambda(h) + 0] \\
&= \lambda(h) \sum_{k=1}^{M} \theta[(\rho_k^*)^2] + \sum_{k=1}^{M} (\rho_k^* - \rho_k^h)^2 \\
&= M^{\lambda(h)}[\rho_1^*, \dots, \rho_M^*] \le M^{\lambda(h)}[\bar{\rho}_1, \dots, \bar{\rho}_M] \\
&= \lambda(h) \sum_{k=1}^{M} \theta(\bar{\rho}_k^2) + \sum_{k=1}^{M} (\bar{\rho}_k - \rho_k^h)^2 \\
&\le \lambda(h) \|\bar{A}^+\|_*^2 + \|\bar{A} - A_h\|^2 \\
&\le \lambda(h) \|\bar{A}^+\|_*^2 + h^2
\end{aligned}$$

It follows that

$$\sum_{k=1}^{M} \theta[(\rho_k^*)^2] \le \|\bar{A}^+\|_*^2 \tag{21}$$

By means of the orthogonal matrices $\bar{U}$ and $\bar{V}$ arising from the singular value decomposition of the matrix $\bar{A}$ and the diagonal matrix

$$D_*^h \equiv \operatorname{diag}(\rho_1^*, \dots, \rho_M^*) \in \mathfrak{A}$$

we can define the matrix

$$A_*^h \equiv \bar{U} D_*^h \bar{V}^{\mathrm{T}}$$

Notice that the matrix D_*^h consists of all singular values of the matrix A_*^h. From formula (9) with $\lambda = \lambda(h) - 0 = \lambda_t - 0$ it follows that

$$\rho_k^* = \rho_k^h x_k(\lambda_{t+1}) > 0$$

for $k = 1, \dots, t(h)$ and

$$\rho_k^* = 0$$

for $k = t(h) + 1, \dots, M$. Hence, $\mathrm{Rg} A_*^h = t(h)$.

Inequalities (20) and (21) can be rewritten as

$$\begin{aligned}
\|A_*^h - \bar{A}\|^2 &= \|\bar{U}(D_*^h - \bar{D})\bar{V}^{\mathrm{T}}\|^2 = \|D_*^h - \bar{D}\|^2 \\
&= \sum_{k=1}^{M} (\rho_k^* - \bar{\rho}_k)^2 \le (3h)^2 \\
\|(A_*^h)^+\|_*^2 &= \sum_{k=1}^{M} \theta[(\rho_k^*)^2] \le \|\bar{A}^+\|_*^2
\end{aligned}$$

The resulting inequalities imply the conditions of Corollary 2.5 and lead to the equality $t(h) = \mathrm{Rg} A_*^h = \mathrm{Rg} \bar{A}$, which is valid for $0 < h < \|\bar{A}^*\|_*^{-1}/3$. Thus, the theorem is completely proved. □

Notice that, from the proof of Theorem 3, it follows that the equality $\lambda(h) = \lambda_1$ is impossible for $0 < h < h_1(\bar{A})$, since otherwise we would have

$\mathrm{Rg}\bar{A} = t(h) = 0$, giving $\bar{A} = 0$. But the preceding conditions $\|A_h\| > h \geq \|\bar{A} - A_h\|$ exclude this case from our consideration. Thus, for all sufficiently small h condition (16) is satisfied for a positive number $t(h)$.

Theorem 3 gives the guidelines for the stable determination of the rank of the matrix $\bar{A}$ (as $h \to 0$) from the available approximate matrix A_h. The essence of the new method is to seek a solution $\lambda(h) > 0$ to equation (14) with further recovery of a number $t(h)$ from condition (16) by comparing the numbers $\lambda_k = 27(\rho_k^h)^4/16$, $k = 1, \ldots, M$, $\lambda_{M+1} = 0$ and $\lambda_0 = +\infty$ with $\lambda(h)$.

As mentioned above, in Theorems 1 and 2 we outline a way of constructing the matrix of the **m.p.m.** method provided that equation (14) possesses an ordinary solution $\lambda(h) > 0$. When this is not the case, we refer to an analogue of the **m.m.p.** method matrix by means of which it is possible to obtain stable approximations to a normal pseudosolution $\bar{z}$ of SLAE (1.1). Theorem 3 with $\lambda(h) = \lambda_{t(h)+1}$ may be of help in achieving this aim. We recommend readers to look through its contents once again.

So, let equation (14) have a generalized solution $\lambda(h) = \lambda_{t(h)+1}$, where $t(h) = \mathrm{Rg}\bar{A}$ for $0 < h < h_1(\bar{A})$. By means of the numbers ρ_k^* taken from the proof of Theorem 3, namely

$$\rho_k^* = \begin{cases} \rho_k^h x_k(\lambda_{t(h)+1}) & k = 1, \ldots, t(h) \\ 0 & k = t(h)+1, \ldots, M \end{cases}$$

we form the matrix $A(h) = U_h D(h) V_h^{\mathrm{T}}$, where

$$D(h) \equiv \mathrm{diag}(\rho_1^*, \ldots, \rho_M^*) \in \mathfrak{A}$$

For any such matrix

$$A^+(h) = V_h D^+(h) U_h^{\mathrm{T}}$$

with $D^+(h) = \mathrm{diag}[\theta(\rho_1^*), \ldots, \theta(\rho_M^*)] \in \mathfrak{A}^*$ and $\hat{z}_\eta = A^+(h)u_\sigma$ will be taken as the needed element.

Theorem 4 *Approximate solutions $\hat{z}_\eta$ converge to $\bar{z}$ as $\eta = (h, \sigma) \to 0$. The following estimate for the rate of convergence*

$$\|\hat{z}_\eta - \bar{z}\| \leq \|\bar{A}^+\|_* \left(\sigma + \frac{3h\|\bar{A}^+\|_*\|\bar{u}\|}{(1 - 3h\|\bar{A}^+\|_*)^3}\right)$$

holds for $0 < h < h_1(\bar{A})$.

Proof. On the basis of estimate (19) we derive the chain of relations

$$\begin{aligned} \|A(h) - A_h\|^2 &= \|U_h[D(h) - D_h]V_h^{\mathrm{T}}\|^2 = \|D(h) - D_h\|^2 \\ &= \sum_{k=1}^{M} (\rho_k^* - \rho_k^h)^2 = \beta[\lambda(h) + 0] \leq 4h^2 \end{aligned}$$

yielding

$$\|A(h) - \bar{A}\| \leq \|A(h) - A_h\| + \|A_h - \bar{A}\| \leq 3h \tag{22}$$

Then, by virtue of inequality (21) and equality (2.9), the relations

$$\|A^+(h)\|_*^2 = \sum_{k=1}^{M} \theta[(\rho_k^*)^2] \le \|\bar{A}^+\|_*^2 \qquad (23)$$

should occur.

Arguing as in the proof of Theorem 2.6 we deduce from inequalities (22)–(23) that

$$\|A^+(h) - \bar{A}^+\|_* \le \frac{3h\|\bar{A}^+\|_*^2}{(1-3h\|\bar{A}^+\|_*)^3}$$

for $0 < h < h_1(\bar{A})$. This implies, by analogy with relation (2.16), the sharp estimate for the rate of convergence:

$$\begin{aligned} \|z_\eta - \bar{z}\| &\le \|\bar{A}^+\|_*\sigma + \|A^+(h) - \bar{A}^+\|_* \cdot \|\bar{u}\| \\ &\le \|\bar{A}^+\|_*\left(\sigma + \frac{3h\|\bar{A}^+\|_*\|\bar{u}\|}{(1-3h\|\bar{A}^+\|_*)^3}\right) \end{aligned}$$

Thus, the theorem is completely proved. □

This theorem substantiates that the matrix of the **m.m.p** method for constructing approximations to $\bar{z}$ can be replaced by the matrix $A(h)$. The approximate solution $\hat{z}_\eta$ related to that matrix is of optimal accuracy order $O(h+\sigma)$ disregarding to the fact whether system (1.1) is compatible or not. However, the matrix $A(h)$ does not satisfy the inequality $\|A - A_h\| \le h$.

The second procedure works for seeking a solution of problem (1) and consists of finding for a fixed $r, 1 \le r \le M$, numbers $\breve{\rho}_1, \dots, \breve{\rho}_r > 0$ such that

$$\begin{aligned} \sum_{k=1}^{r} \breve{\rho}_k^{-2} = \inf \Bigg\{ &\sum_{k=1}^{r} \rho_k^{-2} : \rho_1, \dots, \rho_r > 0 \\ &\sum_{k=1}^{r} (\rho_k - \rho_k^h)^2 + \sum_{k=r+1}^{M} (\rho_k^h)^2 = h^2 \Bigg\} \end{aligned} \qquad (24)$$

We spoke above about the Lagrange function. While solving problem (24) we refer to that function once again:

$$\begin{aligned} \mathcal{L}^\mu(\rho_1, \dots, \rho_r) = \mu \sum_{k=1}^{r} \rho_k^{-2} + \sum_{k=1}^{r} (\rho_k - \rho_k^h)^2 \\ \mu \ge 0 \qquad \rho_1, \dots, \rho_r > 0 \end{aligned} \qquad (25)$$

Lemma 4 *Let $r = Rg\bar{A}$ and $0 < h < h_0(\bar{A})$. Then problem (24) has a unique solution $\breve{\rho}_1, \dots, \breve{\rho}_r > 0$ possessing the monotonicity property $\breve{\rho}_1 \ge \dots \ge \breve{\rho}_r > 0$. The numbers $\breve{\rho}_k$, $k = 1, \dots, r$, are determined by the relation $\breve{\rho}_k = \rho_k^h x_k[\mu(h)]$, where $x_k(\mu)$ is a positive root to the equation*

$$x^4 - x^3 = \mu(\rho_k^h)^{-4} \qquad \mu \ge 0 \qquad (26)$$

and the number $\mu(h) \geq 0$ is a unique root to the equation

$$\varepsilon(\mu) \equiv \sum_{k=1}^{r} (\rho_k^h)^2 [x_k(\mu) - 1]^2 + \sum_{k=r+1}^{M} (\rho_k^h)^2 = h^2 \tag{27}$$

where the function $\varepsilon(\mu)$ defined for $\mu \geq 0$ is continuous and monotonically increasing.

Proof. The necessary conditions for the extremum of Lagrange's function (25) yield $\rho_k^3(\rho_k - \rho_k^h) = \mu$ for $k = 1, \dots, r$. Upon replacing $x_k = \rho_k/\rho_k^h$ the last equation reduces to an equation similar to (26). Here we make use of Corollary 2.4, namely the inequality $\mathrm{Rg}A_h \geq \mathrm{Rg}\bar{A} = r$, which implies $\rho_k^h > 0$ for $k = 1, \dots, r$. As stated above, equation (26) has a unique positive solution $x_k(\mu) \geq 1$ for $\mu \geq 0$. Therefore, function (25) possesses the unique stationary point $\rho_k(\mu) = \rho_k^h x_k(\mu)$, $k = 1, \dots, r$, realizing its minimum, since

$$d^2 \mathcal{L}^\mu [\rho_1(\mu), \dots, \rho_r(\mu)] = 2 \sum_{k=1}^{r} [1 + 3\mu\rho_k^{-4}(\mu)](d\rho_k)^2 > 0$$

Next, the Lagrange multiplier $\mu \geq 0$ is sought from the connection condition of problem (24) taking the form of equation (27) for $\rho_k = \rho_k(\mu)$. It should be noted that the function $\varepsilon(\mu)$ defined for $\mu \geq 0$ is monotonically increasing and continuous in light of similar properties of functions $x_k(\mu)$, $k = 1, \dots, r$. Clearly, (26) gives $x_k(0) = 1$ and $x_k(+\infty) = +\infty$, so that $\varepsilon(+\infty) = +\infty$ and

$$\varepsilon(0) = \sum_{k=r+1}^{M} (\rho_k^h)^2 \leq \sum_{k=r+1}^{M} (\rho_k^h)^2 + \sum_{k=1}^{r} (\bar{\rho}_k - \rho_k^h)^2 \leq \|\bar{A} - A_h\|^2 \leq h^2$$

The properties of the function $\varepsilon(\mu)$ established above provide the unique solvability of equation (27). Therefore, the numbers $\breve{\rho}_1, \dots, \breve{\rho}_r, \mu(h)$ represent a saddle point of Lagrange's function (25) and, consequently, the values $\breve{\rho}_1, \dots, \breve{\rho}_r$ constitute a solution of problem (24).

It remains to show that this solution is unique. Assume to the contrary that there were two solutions of problem (24): the preceding one and, say $\tilde{\rho}_1, \dots, \tilde{\rho}_r > 0$. Then

$$\sum_{k=1}^{r} \tilde{\rho}_k^{-2} = \sum_{k=1}^{r} \breve{\rho}_k^{-2}$$

$$\sum_{k=1}^{r} (\tilde{\rho}_k - \rho_k^h)^2 = \sum_{k=1}^{r} (\breve{\rho}_k - \rho_k^h)^2 = h^2 - \sum_{k=r+1}^{M} (\rho_k^h)^2$$

whence the equality $\mathcal{L}^{\mu(h)}(\tilde{\rho}_1, \dots, \tilde{\rho}_r) = \mathcal{L}^{\mu(h)}(\breve{\rho}_1, \dots, \breve{\rho}_r)$ immediately follows. On the other hand, since function (25) possesses a unique global minima, the strict inequality

$$\mathcal{L}^{\mu(h)}(\breve{\rho}_1, \dots, \breve{\rho}_r) < \mathcal{L}^{\mu(h)}(\tilde{\rho}_1, \dots, \tilde{\rho}_r)$$

should occur. The obtained contradiction proves the uniqueness of the problem (24) solution. Its monotonicity can be shown in just the same way as we did in Lemma 2. □

A precise relationship between solutions of problems (1) and (24) is of great interest. The following proposition gives an answer to this question.

Theorem 5 *If the condition $\|A_h\| > h$ holds for $0 \le h < h_0(\bar{A})$, then problem* (1) *has the unique solution $(\hat{\rho}_1, \dots, \hat{\rho}_M) = (\check{\rho}_1, \dots, \check{\rho}_r, 0, \dots, 0)$, where $(\check{\rho}_1, \dots, \check{\rho}_r)$ is a solution of problem* (24) *from Lemma* 4.

Proof. By Theorem 2.4, for $0 \le h < h_0(\bar{A})$ the rank of any matrix of the **m.p.m.** method is exactly r. So, the rank of the matrix $\widehat{A}_h$ from Theorem 1 is r and, consequently, the rank of the matrix $\widehat{D}_h = \mathrm{diag}(\hat{\rho}_1, \dots, \hat{\rho}_M)$, where $\hat{\rho}_1, \dots, \hat{\rho}_M$ are solutions of problem (1), is the same. Therefore, $\hat{\rho}_k > 0$ for $1 \le k \le r$ and $\hat{\rho}_k = 0$ for $r < k \le M$. It is easily seen from (1) that for such numbers $\hat{\rho}_k$

$$\sum_{k=1}^{M} (\hat{\rho}_k - \rho_k^h)^2 = \sum_{k=1}^{r} (\hat{\rho}_k - \rho_k^h)^2 + \sum_{k=r+1}^{M} (\rho_k^h)^2 = h^2$$

$$\sum_{k=1}^{M} \theta(\hat{\rho}_k^2) = \sum_{k=1}^{r} \hat{\rho}_k^{-2} \le \sum_{k=1}^{r} \check{\rho}_k^{-2}$$

The proof of the last inequality exploits the fact that the numbers

$$(\check{\rho}_1, \dots, \check{\rho}_r, 0, \dots, 0)$$

satisfy the requirements of the conditional extremal problem (1) (see problem (24) and Lemma 4). We may infer from these conditions that the numbers $\hat{\rho}_1, \dots, \hat{\rho}_r$ constitute a solution of problem (24). In view of the uniqueness of this solution, the equalities $\hat{\rho}_k = \check{\rho}_k$, $k = 1, \dots, r$, should be valid. Thus, the theorem is completely proved. □

For $0 < h < h_0(\bar{A})$, when determining the number r according to Theorem 3, one can find a solution of problem (1) by the formulae from Lemma 4.

Theorem 5 implies a profound result.

Theorem 6 *Under the conditions $0 \le h < h_0(\bar{A})$ and $\|A_h\| > h$ any matrix of the* **m.p.m.** *method has the same singular values $\hat{\rho}_1, \dots, \hat{\rho}_M$ being the unique solution of problem* (1).

Proof. Let an arbitrary matrix $\widetilde{A}_h$ of the **m.p.m.** method give a solution of problem (2.1). We denote its singular values by $\tilde{\rho}_1 \ge \dots \ge \tilde{\rho}_M \ge 0$. We know from Theorem 1 that the matrix $\widehat{A}_h$ with singular values $\hat{\rho}_1 \ge \dots \ge \hat{\rho}_M \ge 0$ represents a solution of problem (2.1). Therefore, combining (2.1),

(2.8) and (2.9) we establish the relations

$$\|\widetilde{A}_h^+\|_*^2 = \sum_{k=1}^{M} \theta(\tilde{\rho}_k^2) = \|\hat{A}_h^+\|_*^2 = \sum_{k=1}^{M} \theta(\hat{\rho}_k^2) \tag{28}$$

$$\|\widetilde{A}_h - A_h\|^2 = h^2 \geq \sum_{k=1}^{M} (\tilde{\rho}_k - \rho_k^h)^2 \tag{29}$$

It is clear from (29) that the numbers $\tilde{\rho}_k$, $k = 1, \dots, M$, satisfy the requirements of the conditional extremal problem (2) which is equivalent by Lemma 1 to problem (1). It follows from (28) that the greatest lower bound in problem (2) and, consequently, in problem (1) can be attained for exactly the same numbers and, therefore, the collection $\tilde{\rho}_1, \dots, \tilde{\rho}_M$ is just a solution of problem (1). Under the conditions of the theorem, this solution is unique and possesses the property $\tilde{\rho}_k = \hat{\rho}_k$, $k = 1, \dots, M$. With this, we finish the proof of the theorem. □

We hope that several examples add interest and help in understanding.

Example 1 (The search of singular numbers of the matrix of the m.p.m. method) Let the singular values of the exact matrix $\bar{A}$ be known: $\rho_1 = 1$ and $\rho_k = 0$, $k = 2, \dots, M$, then the singular values of the approximate matrix A_h may be taken to be $\rho_1^h = 1$, $\rho_2^h = h_2, \dots, \rho_M^h = h_M$ with $h_k \geq 0$ and $h_2^2 + \dots + h_M^2 \leq h^2$. By Theorem 5, one can state that for $0 < h < 1/2$ the singular values $\hat{\rho}_k$ of the matrix of the **m.p.m.** method are equal to 0 for $k = 2, \dots, M$. Since the numbers $\hat{\rho}_k$, $k = 1, \dots, M$, obey the constraints of problem (1), the following relation occurs:

$$\sum_{k=1}^{M} (\hat{\rho}_k - \rho_k^h)^2 = (\hat{\rho}_1 - 1)^2 + \sum_{k=2}^{M} h_k^2 = h^2$$

With this, solving problem (1) is equivalent to making the proper choice among two numbers satisfying the last equality, the one for which the value $\theta(\hat{\rho}_1^2) = \hat{\rho}_1^{-2}$ is less. Therefore,

$$\hat{\rho}_1 = 1 + \left[h^2 - \sum_{k=2}^{M} h_k^2\right]^{1/2}$$

Example 2 (Stable determination of the rank of the matrix $\bar{A}$) Let

$$\bar{A} = \begin{pmatrix} 1 & 0 \\ 0 & 0 \end{pmatrix} \qquad A_h = \begin{pmatrix} 1 & 0 \\ 0 & h \end{pmatrix} \qquad 0 < h < \frac{1}{2}$$

In this case, $\rho_1^h = 1$, $\rho_2^h = h$, $\lambda_1 = 27/16$ and $\lambda_2 = 27h^4/16$. One is to calculate the function $\beta(\lambda)$ in complete agreement with Lemma 3. A final

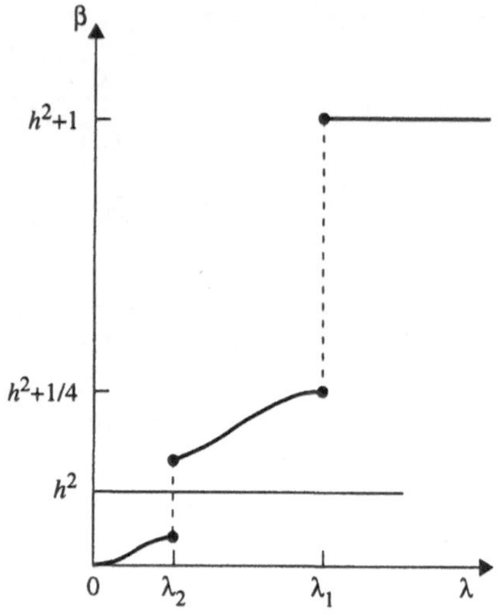

Figure 6.4 Illustration to Example 2.

result is:

$$\beta(\lambda) = [\rho_1(\lambda) - \rho_1^h]^2 + [\rho_2(\lambda) - \rho_2^h]^2$$
$$= \begin{cases} [x_1(\lambda) - 1]^2 + h^2[x_2(\lambda) - 1]^2 & 0 \le \lambda \le \lambda_2 \\ [x_1(\lambda) - 1]^2 + h^2 & \lambda_2 \le \lambda \le \lambda_1 \\ 1 + h^2 & \lambda \ge \lambda_1 \end{cases}$$

The graph of this function is depicted in Fig. 6.4.

In the given example equation (14) has a generalized solution $\lambda(h) = \lambda_2$, since (13) implies the inequalities

$$\beta(\lambda_2 - 0) = \tfrac{h^2}{4} + [x_1(\lambda_2) - 1]^2 \le \tfrac{h^2}{4} + \lambda_2^2 = \tfrac{h^2}{4} + \left(\tfrac{27h^4}{16}\right)^2 < h^2$$

$$\beta(\lambda_2 + 0) = \beta(\lambda_2 - 0) + \tfrac{3h^2}{4} > h^2$$

basing on the estimate $0 \le 1 - x_1(\lambda) \le \lambda$, which is an immediate implication of the equation $x_1^4 - x_1^3 = \lambda$ on the segment $1 \le x_1 \le 3/2$. From here it follows that $\lambda_2 \le \lambda(h) < \lambda_1$, yielding, by Theorem 3, $\text{Rg}\bar{A} = t(h) = 1$ for $0 < h < \frac{1}{2}$.

6.4 The m.p.m. principle

An alternative approach to solving problems (2.1) and (2.4) during the course of the **m.p.m.** method is possible. This involves quite different ideas about employing the method of undetermined Lagrange multipliers with the use of a special functional

$$M^\lambda[A] = \lambda\|A^+\|_*^2 + \|A - A_h\|^2, \qquad \lambda > 0 \qquad A \in \mathfrak{A}_0 \tag{1}$$

and the accompanying extremal problem in which for a fixed $\lambda > 0$ it is required to find a matrix extremal $A_\lambda \equiv A_{h\lambda} \in \mathfrak{A}_0$ such that

$$M^\lambda[A_\lambda] = \inf\left\{M^\lambda[A] \colon A \in \mathfrak{A}_0\right\} \equiv \varphi(\lambda) \tag{2}$$

Theorem 1 *For every $\lambda > 0$ problem (2) is solvable.*

Proof. Let $\{A_n\}$ be a minimizing sequence for problem (2) if $\lambda > 0$ is held fixed: $A_n \in \mathfrak{A}_0$ and $M^\lambda[A_n] \to \varphi(\lambda)$ as $n \to \infty$. Then there exists a number $\varepsilon(n_0)$ such that for $n \geq n_0$

$$\begin{aligned} \|A_n\| &\leq \|A_h\| + \|A_n - A_h\| \leq \|A_h\| + \{M^\lambda[A_n]\}^{1/2} \\ &\leq \|A_h\| + \{\varphi(\lambda) + \varepsilon(n_0)\}^{1/2} \equiv M_1 = \text{const} \\ \|A_n^+\|_* &\leq \{M^\lambda[A_n]/\lambda\}^{1/2} \leq \{[\varphi(\lambda) + \varepsilon(n_0)]/\lambda\}^{1/2} \equiv M_2 = \text{const} \end{aligned}$$

Due to Corollary 2.1 and Lemma 2.1, these inequalities imply the existence of a subsequence $\{A_{n_l}\}_{l=1}^\infty$ such that $A_{n_l} \to A_0 \in \mathfrak{A}_0$ and $A_{n_l}^+ \to A_0^+$ as $l \to \infty$. In this line,

$$\begin{aligned} \lim_{l\to\infty} M^\lambda[A_{n_l}] &= \lim_{l\to\infty} \left\{\lambda\|A_{n_l}^+\|_*^2 + \|A_{n_l} - A_h\|^2\right\} \\ &= \lambda\|A_0^+\|_*^2 + \|A_0 - A_h\|^2 = \varphi(\lambda) \\ &= \inf\left\{M^\lambda[A] \colon A \in \mathfrak{A}_0\right\} \end{aligned}$$

This provides support for making the final decision that for given λ the matrix A_0 can be adopted as a solution of problem (2). □

In what follows we will show that problem (2) may be nonuniquely solvable (at least for some numbers λ).

For later use, we define for a fixed h two multifunctions.

$$\beta(\lambda) \equiv \|A_\lambda - A_h\|^2 \qquad\qquad \gamma(\lambda) \equiv \|A_\lambda^+\|_*^2 \tag{3}$$

Clearly, they are similar to the auxiliary functions $\beta(\alpha)$ and $\gamma(\alpha)$ which have been under consideration in Sections 2.6 and 3.2in Chapter 2 and 3 respectively. To take this as basis, we must adopt the techniques similar to those of Section 2.6 in establishing the fundamental properties of functions (3). In particular, the following assertion is valid for $\beta(\lambda)$.

Lemma 1 *The function $\beta(\lambda)$ is monotonically nondecreasing and, for $\lambda > 0$, single-valued everywhere except at most a countable set of discontinuity points of the first kind being multivalence ones. When λ_0 is a discontinuity point of the function $\beta(\lambda)$, the set of all extremals of functional (1) for this value of λ contains at least two matrices $A_{\lambda_0 \pm 0}$ satisfying the relation $\beta(\lambda_0 \pm 0) = \|A_{\lambda_0\pm 0} - A_h\|^2$. If, in addition, $0 \in \mathfrak{A}_0$ and $A_h \in \mathfrak{A}_0$, then $\beta(+0) = 0$ and $\beta(+\infty) = \|A_h\|^2$.*

Observe that, in proving the existence of matrices $A_{\lambda_0 \pm 0}$ by analogy with Lemma 2.6.5 in Chapter 2, we will use the results of Lemma 2.1 and Corollary 2.1 rather than certain properties of the stabilizer Ω (the lower semicontinuity in some topology and the compactness of the stabilizer level sets in the same topology).

Lemmas 1 and 2.6.7 imply one interesting result.

Theorem 2 *Let* $0 \in \mathfrak{A}_0$, $A_h \in \mathfrak{A}_0$ *and* $\|A_h\|^2 > Ch^2 > 0$, *where* $C = const > 1$. *Then the equation with a monotone function*

$$\beta(\lambda) = Ch^2 \tag{4}$$

possesses a solution $\lambda(h) > 0$.

By following this theorem, we take the value $\lambda = \lambda(h)$ recovered from (4) and choose from the set of the problem (2) solutions for $\lambda = \lambda(h)$ a matrix

$$A_{\lambda(h)} \equiv A_{\lambda(h)-0}$$

such that

$$\beta[\lambda(h) - 0] = \|A_{\lambda(h)} - A_h\|^2 \leq Ch^2 \tag{5}$$

Here we have taken (4) into account. It is easily seen that the matrix $A_{\lambda(h)}$ found in such a way is fitted into the framework of the **g.p.d.** algorithm for solving the compatible operator equation $B(A) = \bar{A}$ on the set $\mathfrak{A}_0$ consisting of matrices A. In the last equation, B is the identity operator from $\mathfrak{A}$ into $\mathfrak{A}$ and the right-hand side $\bar{A}$ is specified approximately, since A_h is involved instead of $\bar{A}$. The value $\|A^+\|_*^2$ will play the role of the functional Ω.

The next theorem is deduced from the results of Section 3.6 in Chapter 3.

Theorem 3 *The estimate*

$$\|A^+_{\lambda(h)}\|_*^2 \leq \frac{C}{C-1}\|\bar{A}^+\|_*^2 \tag{6}$$

is true.

Theorem 3 implies

Theorem 4 *Under the conditions of Theorem* 2

$$A^+_{\lambda(h)} \to \bar{A}^+$$

as $h \to 0$.

Proof. Indeed, it is evident from (5) that $A_{\lambda(h)} \to \bar{A}$ as $h \to 0$. By Lemma 2.1, the latter along with estimate (6) implies the desired convergence. □

We begin by defining the elements $\hat{z}_\eta = A^+_{\lambda(h)} u_\sigma$. Using inequality (2.6) and the convergence established in Theorem 4, one can prove the following assertion.

Theorem 5 *Let for every $h > 0$ the conditions of Theorem 2 hold. Then $\hat{z}_\eta \to \bar{z}$ as $\eta = (h, \sigma) \to 0$.*

We say that the element $\hat{z}_\eta$ is obtained by the **m.p.m. principle**. The matrix $A_{\lambda(h)}$ is called the **matrix of the m.p.m. principle**.

In the case when $\mathfrak{A}_0 = \mathfrak{A}$ the **m.p.m.** principle may be simplified to a great extent and its matrix can be calculated directly. This peculiarity needs investigation.

Theorem 6 *For $\mathfrak{A}_0 = \mathfrak{A}$ all of the solution to problem (2) for a given $\lambda > 0$ take the form*

$$\widehat{A}_\lambda = \widehat{U}\widehat{D}_\lambda \widehat{V}^{\mathrm{T}}$$

Here, the diagonal matrix $\widehat{D}_\lambda = diag[\rho_1(\lambda), \ldots, \rho_M(\lambda)] \in \mathfrak{A}$ contains the singular values of the matrix $\widehat{A}_\lambda$ specified by formula (3.9) and both orthogonal matrices $\widehat{U} \in \mathfrak{A}_m$ and $\widehat{V} \in \mathfrak{A}_n$ satisfy the equality

$$\|\widehat{U}\widehat{D}_\lambda \widehat{V}^{\mathrm{T}} - A_h\| = \|\widehat{D}_\lambda - D_h\|$$

In particular, the matrices U_h and V_h entering the singular decomposition of A_h may be taken to be $\widehat{U}$ and $\widehat{V}$.

Proof. We first show that the matrix $\widehat{A}_\lambda$ mentioned in the theorem gives a solution of problem (2). By the restriction imposed on the matrices $\widehat{U}$ and $\widehat{V}$ and equality (2.9), we arrive at the chain of relations (see (3.8)):

$$\begin{aligned} M^\lambda[\widehat{A}_\lambda] &= \lambda\|\widehat{A}_\lambda^+\|_*^2 + \|\widehat{A}_\lambda - A_h\|^2 \\ &= \lambda\|\widehat{D}_\lambda^+\|_*^2 + \|\widehat{D}_\lambda - D_h\|^2 \\ &= \lambda\sum_{k=1}^{M}\theta[\rho_k^2(\lambda)] + \sum_{k=1}^{M}[\rho_k(\lambda) - \rho_k^h]^2 \\ &= \mathcal{L}^\lambda[\rho_1(\lambda), \ldots, \rho_M(\lambda)] \end{aligned} \tag{7}$$

Since, due to Lemma 3.2, only the numbers $\rho_k(\lambda)$, $k = 1, \ldots, M$, permit to reach the greatest lower bound of (3.8), we might have

$$M^\lambda[\widehat{A}_\lambda] = \mathcal{L}^\lambda[\rho_1(\lambda), \ldots, \rho_M(\lambda)] < \mathcal{L}^\lambda[\rho_1, \ldots, \rho_M]$$

for any collection of numbers $(\rho_1, \ldots, \rho_M)$, $\rho_1 \geq \cdots \geq \rho_M \geq 0$, other than $[\rho_1(\lambda), \ldots, \rho_M(\lambda)]$. Taking into account the form (3.7) of the function $\mathcal{L}^\lambda$ we derive with the aid of relations (2.8)–(2.9) that

$$\begin{aligned} M^\lambda[\widehat{A}_\lambda] &< \mathcal{L}^\lambda(\rho_1, \ldots, \rho_M) \\ &= \lambda\sum_{k=1}^{M}\theta(\rho_k^2) + \sum_{k=1}^{M}(\rho_k - \rho_k^h)^2 \\ &\leq \lambda\|A^+\|_*^2 + \|A - A_h\|^2 = M^\lambda[A] \end{aligned} \tag{8}$$

which makes sense for any matrix A with singular values $\rho_1, \ldots, \rho_M$. Since they may be taken arbitrarily, the inequality

$$M^\lambda[\widehat{A}_\lambda] < M^\lambda[A]$$

is valid for any matrix $A \in \mathfrak{A}$ whose singular values differ from the collection $[\rho_1(\lambda), \ldots, \rho_M(\lambda)]$. This means that $\widehat{A}_\lambda$ is just a solution of problem (2) for $\mathfrak{A}_0 = \mathfrak{A}$.

It remains to prove that problem (2) has no solutions other than the matrices $\widehat{A}_\lambda$. We claim that, in view of relations (7)–(8), any solution of problem (2) with $\mathfrak{A}_0 = \mathfrak{A}$ possesses the same singular values $\rho_k(\lambda)$, $k = 1, \ldots, M$. Indeed, assuming that problem (2) admits a solution

$$A_* = \widehat{U}\widehat{D}_\lambda\widehat{V}^{\mathrm{T}}$$

and keeping in mind that the conditions of the theorem may be violated for the orthogonal matrices $\widehat{U}$ and $\widehat{V}$ thus obtained, we deduce from (2.8) that

$$\|A_* - A_h\| = \|\widehat{U}\widehat{D}_\lambda\widehat{V}^{\mathrm{T}} - A_h\| > \|\widehat{D}_\lambda - D_h\|$$

Putting these together with (7) we get the estimate

$$\begin{aligned} M^\lambda[\widehat{A}_\lambda] &= \lambda\|\widehat{D}_\lambda^+\|_*^2 + \|\widehat{D}_\lambda - D_h\|^2 \\ &< \lambda\|\widehat{V}\widehat{D}_\lambda^+\widehat{U}^{\mathrm{T}}\|_*^2 + \|A_* - A_h\|^2 \\ &= \lambda\|A_*^+\|_*^2 + \|A_* - A_h\|^2 = M^\lambda[A_*] \end{aligned}$$

thereby clarifying that the matrix A_* does not solve problem (2) and justifying the assertion of the theorem. □

We infer from Theorem 6 and formula (3.9) that for every $\lambda = \lambda_k = 27(\rho_k^h)^4/16$, $k = 1, \ldots, M$, problem (2) with $\mathfrak{A}_0 = \mathfrak{A}$ can have at least two solutions with distinct kth singular values $\rho_k(\lambda_k) = 3\rho_k^h/2$ and $\rho_k(\lambda_k) = 0$.

Theorem 6 provides an excellent background for a more careful analysis of the function $\beta(\lambda)$. Indeed, applying successively Theorem 6 and Lemma 3.3 yields the representation

$$\beta(\lambda) = \sum_{k=1}^{M} [\rho_k(\lambda) - \rho_k^h]^2$$

which clarifies that the function $\beta(\lambda)$ so constructed coincides with the function involved in (3.12). Therefore, when $\mathfrak{A}_0 = \mathfrak{A}$ the function $\beta(\lambda)$ possesses the extra properties from Lemma 3.3 in addition to those from Lemma 1. The function $\beta(\lambda)$ has a finite number of discontinuity points $\lambda_1 \geq \cdots \geq \lambda_M \geq 0$, where $\lambda_k = 27(\rho_k^h)^4/16$ ($k = 1, \ldots, M$), monotonically increases for $0 \leq \lambda \leq \lambda_1$ and takes the value $\beta(\lambda) = \beta(+\infty) = \|A_h\|^2$ for $\lambda \geq \lambda_1$.

These properties provide the uniqueness of the solution $\lambda(h) > 0$ to the equation $\beta(\lambda) = h^2$ for $\|A_h\| > h > 0$. The number $\lambda(h)$ chosen in such a way gives a solution to equation (3.14), whose use permits us to construct in accordance with Theorem 6 the following matrices

$$\widehat{A}_{\lambda(h)} = \widehat{U}\widehat{D}_{\lambda(h)}\widehat{V}^{\mathrm{T}}$$

The next assertion is a final result of joint use of Theorems 3.1, 3.2, 3.5 and 6.

Theorem 7 *Let the condition* $\|A_h\| > h$ *hold for* $0 < h < h_0(\bar{A})$ *and let* $\lambda(h)$ *be an ordinary solution to equation* (4) *with* $C = 1$. *Then* $\widehat{A}_{\lambda(h)}$ *represent the matrices of the* **m.p.m.** *method and the* **m.p.m.** *principle simultaneously.*

Thus, we have established an interconnection between the **m.p.m.** method and **m.p.m.** principle similar to that of Section 3.5 in Chapter 3 between the **g.m.d.** and **g.p.d.** algorithms.

Observe that the matrix $\widehat{D}_{\lambda(h)}$ contains the singular values $\hat{\rho}_k = \rho_k[\lambda(h)]$ (see Theorem 3.2) and the matrices U_h and V_h can be viewed as $\widehat{U}$ and $\widehat{V}$.

6.5 Optimal properties of the m.p.m. method

We investigate the optimal accuracy properties of the **m.p.m.** method by comparing with others on classes of systems with approximate data.

For simplicity, let $\mathfrak{A}_0 = \mathfrak{A}$.

Definition 1 *By a* **standard method of the approximate determination of a normal pseudosolution** *of SLAE* (1.1) *we mean any mapping* P *associating with each collection* $(A_h, u_\sigma, h, \sigma)$ *of the problem* (1.1) *approximate data, where* $0 \le h < H_0$ *and* $0 \le \sigma \le \sigma_0$, *an element* $z_\eta = P(A_h, u_\sigma, h, \sigma) \in \mathbf{R}^n$, *which can be expressed by*

$$z_\eta = V_h F(D_h, \alpha_\eta) U_h^{\mathrm{T}} u_\sigma$$

Here $\alpha_\eta \ge 0$ *is the regularization parameter chosen in a special manner and depending, in general, on* $(A_h, u_\sigma, h, \sigma)$; *the matrix* $F(D_h, \alpha_\eta) \in \mathfrak{A}^*$ *is of the form*

$$F(D_h, \alpha_\eta) = diag\Big[f_1(\rho_1^h, \alpha_\eta), \dots, f_M(\rho_M^h, \alpha_\eta)\Big]$$

where the functions $f_k(\rho, \alpha)$ $(k = 1, \dots, M)$ *defined for* $\rho \ge 0$ *and* $\alpha \ge 0$ *are measurable with respect to* ρ *for every fixed* $\alpha \ge 0$ *and bounded on all half-open intervals of the form* $\rho \ge \varepsilon$ *for every* $\varepsilon > 0$.

In partucular, Definition 1 covers the situations in which dependence upon α_η is missing. For example, in applications of the least squares method, we might have

$$z_\eta = A_h^+ u_\sigma = V_h D_h^+ U_h^{\mathrm{T}} u_\sigma$$

and $f_k = \theta(\rho_k^h)$ $(k = 1, \dots, M)$. The standard methods include also algorithms for solving SLAEs such as the regularization method:

$$z_\eta = (\alpha_\eta I + A_h^{\mathrm{T}} A_h)^{-1} A_h^{\mathrm{T}} u_\sigma = V_h(\alpha_\eta I + D_h^{\mathrm{T}} D_h)^{-1} D_h^{\mathrm{T}} U_h^{\mathrm{T}} u_\sigma$$
$$f_k = [\alpha_\eta + (\rho_k^h)^2]^{-1} \rho_k^h \qquad \alpha_\eta > 0$$

the method from the original works of Gilyazov and Morozov (1984) or of Vaĭnikko and Veretennikov (1986):

$$\begin{aligned} z_\eta &= (\alpha_\eta I + A_h^{\mathrm{T}} A_h)^{-1} A_h^{\mathrm{T}} A_h A_h^{\mathrm{T}} (\alpha_\eta E + A_h A_h^{\mathrm{T}})^{-1} u_\sigma \\ &= V_h(\alpha_\eta I + D_h^{\mathrm{T}} D_h)^{-1} D_h^{\mathrm{T}} D_h D_h^{\mathrm{T}} (\alpha_\eta E + D_h D_h^{\mathrm{T}}) U_h^{\mathrm{T}} u_\sigma \\ f_k &= [\alpha_\eta + (\rho_k^h)^2]^{-2} (\rho_k^h)^3 \qquad \alpha_\eta > 0 \end{aligned}$$

and some others (see, for example, Buša (1987), Golub (1968), Lawson and Hanson (1974)).

Theorem 1 *The* **m.p.m.** *method for solving the fundamental problem of Section* 1 *falls in the category of standard methods if*

$0 \le h < H_0(\bar{A}) = \text{const}$.

Proof. Under the agreement of Section 1, $\bar{A} \neq 0$. Therefore, there exists a number $H_0(\bar{A}) = \min\{\|\bar{A}^+\|_*^{-1}/2, \|\bar{A}\|/2\}$ such that for $0 \le h < H_0(\bar{A})$ the hypotheses of Theorems 3.6 and 3.2 will be true. According to these theorems, an approximate solution z_η of the fundamental problem we formulated at the very beginning admits by the construction of the **m.p.m.** method the representation

$$z_\eta = \widehat{A}_h^+ u_\sigma = V_h \widehat{D}_h^+ U_h^{\mathrm{T}} u_\sigma \qquad \widehat{D}_h^+ = \mathrm{diag}[\theta(\hat{\rho}_1), \dots, \theta(\hat{\rho}_M)]$$

The explicit form of the functions $f_k(\rho, \alpha)$ from Definition 1 of the **m.p.m.** method is given by

$$f_k = f(\rho, \alpha) = \begin{cases} [\rho X(\alpha, \rho)]^{-1} & \rho > 0 \quad 0 \le \alpha < 27\rho^4/16 \\ 0 & \rho \ge 0 \quad \alpha \ge 27\rho^4/16 \end{cases}$$

Here $X(\alpha, \rho)$ is a positive solution to the equation $X^4 - X^3 = \alpha\rho^{-4}$ (cf. (3.10′)). According to formula (3.9) we find $f(\rho_k^h, \lambda) = \theta[\rho_k(\lambda)]$ and, by Theorem 3.2,

$$\theta(\hat{\rho}_k) = \theta\{\rho_k[\lambda(h)]\} = f[\rho_k^h, \lambda(h)]$$

The number $\lambda(h) > 0$ can uniquely be recovered from equation (3.14) and can be adopted as a suitable regularization parameter, which means that $\alpha_\eta = \lambda(h)$. To complete the proof, we observe that the function $f(\rho, \alpha)$ satisfies the requirements of Definition 1, since the quantity $X(\alpha, \rho)$ depends continuously on its arguments α, ρ for $\alpha \ge 0$ and $\rho > 0$ (see property (4) for a solution to equation (3.10′)). □

We denote by S the class of standard methods. For fixed numbers R_1 and $R_2 > 0$ both symbols

$$\Sigma \equiv \Sigma(R_1, R_2) = \{(\bar{A}, \bar{u}) \colon \|\bar{A}^+\|_* \le R_1, \|\bar{u}\| \le R_2\}$$

will stand for the class of all exact systems having the form (1.1). Further, holding an exact system $(\bar{A}, \bar{u}) \in \Sigma$ fixed, it is reasonable to introduce the class of approximate data corresponding to an accuracy $\eta = (h, \sigma)$

$$\Sigma_\eta \equiv \Sigma_\eta(\bar{A}, \bar{u}) = \{(A_h, u_\sigma) \colon \|A_h - \bar{A}\| \le h, \|u_\sigma - \bar{u}\| \le \sigma\}$$

Of special interest will be the **accuracy characteristic** of a method $P \in S$ on the class $\Sigma(R_1, R_2)$ of all exact systems when available approximate data happen to be from the appropriate class $\Sigma_\eta(\bar{A}, \bar{u})$:

$$\begin{aligned}\Delta(P, \eta) &\equiv \Delta(P, \eta, R_1, R_2)\\ &= \sup_{A_h, u_\sigma, \bar{A}, \bar{u}} \{\|P(A_h, u_\sigma, h, \sigma) - \bar{z}\| \colon \bar{z} = \bar{A}^+\bar{u},\\ &\quad (\bar{A}, \bar{u}) \in \Sigma(R_1, R_2), (A_h, u_\sigma) \in \Sigma_\eta(\bar{A}, \bar{u})\} \qquad (1)\end{aligned}$$

Definition 2 *By the* **optimal accuracy** *of an approximate solution of system* (1.1) *from* $\Sigma(R_1, R_2)$ *on the class of standard methods* S *we mean the quantity*

$$\Delta_0(\eta) \equiv \Delta_0(\eta, R_1, R_2) = \inf\{\Delta(P, \eta) \colon P \in S\}$$

Definition 3 *A method* $P \in S$ *is said to be of* **optimal accuracy order** *if*

$$\overline{\lim}_{\eta \to 0} \frac{\Delta(P, \eta, R_1, R_2)}{\Delta_0(\eta, R_1, R_2)} \le g$$

where a constant $g \ge 1$ *does not depend on* R_1 *and* R_2 *both.*

The following lemma enables one to obtain the lower estimate for the accuracy of any method $P \in S$.

Lemma 1 *For any* $P \in S$ *and any* $\eta = (h, \sigma)$ $(0 \le h < h_0(R_1) \equiv R_1^{-1}/2$, $0 \le \sigma < R_2)$, *the estimate*

$$\begin{aligned}\Delta(P, \eta) \ &\ge\ R_1(\sigma + R_1R_2h)[1 - 2hR_1/q(h)]\\ &\quad \times[1 + q(h)]^{-2}(1 + \sigma/R_2)^{-1} \equiv \nu(h, \sigma)\end{aligned}$$

is valid with $q(h)$ *denoting an arbitrary bounded function for which* $q(h) > 2hR_1$.

Proof. As a first step, we extract from the set $\Sigma(R_1, R_2)$ a subclass Σ_1 of all exact systems with the data $(\bar{A}, \bar{u})$ subject to the condition $\bar{A} = \bar{U}\bar{D}\bar{V}^{\mathrm{T}}$, where $\bar{U} \in \mathfrak{A}_m$, $\bar{V} \in \mathfrak{A}_n$ are orthogonal matrices and $\bar{D} = \mathrm{diag}(\bar{\rho}, 0, \dots, 0) \in \mathfrak{A}$ $(\bar{\rho} > 0)$; $\bar{u} \in \mathrm{Im}\bar{A}$, $\|\bar{u}\| = R_2$. From the set $\Sigma_\eta(\bar{A}, \bar{u})$ corresponding to every collection $(\bar{A}, \bar{u})$ with these properties we choose a subset Σ_η^1 of all approximate data (A_h, u_σ) satisfying the following conditions

$$A_h = \bar{U}D_h\bar{V}^{\mathrm{T}} \qquad D_h = \mathrm{diag}(\tilde{\rho}, 0, \dots, 0) \in \mathfrak{A} \qquad |\bar{\rho} - \tilde{\rho}| \le h$$
$$u_\sigma = t\bar{u} \qquad |t - 1| \le \sigma/R_2$$

The obvious relations

$$\|\bar{A} - A_h\| = \|\bar{D} - D_h\| = |\bar{\rho} - \tilde{\rho}| \leq h \qquad \|\bar{A}^+\|_* = \bar{\rho}^{-1} \leq R_1$$
$$\|\bar{u} - u_\sigma\| = |t-1| \cdot \|\bar{u}\| \leq \sigma$$

occur on the sets Σ_1 and Σ_η^1.

Taking into account the restrictions on $\bar{A}, \bar{u}, A_h, u_\sigma$ we deduce by the definition of standard method that

$$\begin{aligned} \|P(A_h, u_\sigma, h, \sigma) - \bar{A}^+\bar{u}\|^2 &= \|\bar{V}F(D_h, \alpha_\eta)\bar{U}^{\mathrm{T}}(t\bar{u}) - \bar{V}\bar{D}^+\bar{U}^{\mathrm{T}}\bar{u}\|^2 \\ &= \sum_{k=1}^{M} \bar{u}_k^2[f_k(\rho_k^h, \alpha_\eta)t - \theta(\bar{\rho}_k)]^2 \\ &= \bar{u}_1^2[f_1(\tilde{\rho}, \alpha_\eta)t - \bar{\rho}^{-1}]^2 \end{aligned} \tag{2}$$

where $\bar{u}_k$ $(k = 1, \ldots, M)$ are the components of the vector $\bar{U}^{\mathrm{T}}\bar{u} \in \mathbf{R}^m$. It is easily seen that the vector $\bar{U}^{\mathrm{T}}\bar{u}$ takes for now the form $\bar{U}^{\mathrm{T}}\bar{u} = (\bar{u}_1, 0, \ldots, 0)$ with $|\bar{u}_1| = \|\bar{u}\| = R_1$ because $\bar{u} \in \mathrm{Im}\bar{A}$.

In view of this, it follows from (1) and (2) that

$$\begin{aligned} \Delta^2(P, \eta) &= \sup_{\bar{A}, \bar{u}, A_h, u_\sigma} \{\|P(A_h, u_\sigma, h, \sigma) - \bar{A}^+\bar{u}\|^2 : \\ &\qquad (\bar{A}, \bar{u}) \in \Sigma(R_1, R_2), (A_h, u_\sigma) \in \Sigma_\eta(\bar{A}, \bar{u})\} \\ &\geq \sup_{\bar{A}, \bar{u}, A_h, u_\sigma} \{\|P(A_h, u_\sigma, h, \sigma) - \bar{A}^+\bar{u}\|^2 : \\ &\qquad (\bar{A}, \bar{u}) \in \Sigma_1, (A_h, u_\sigma) \in \Sigma_\eta^1\} \\ &= \sup_{\bar{\rho}, \bar{u}_1, \tilde{\rho}, t} \{\bar{u}_1^2[f_1(\tilde{\rho}, \alpha_\eta)t - \bar{\rho}^{-1}]^2 : |\bar{\rho} - \tilde{\rho}| \leq h, \bar{\rho}^{-1} \leq R_1, \\ &\qquad |\bar{u}_1| = R_2, |t-1| \leq \sigma/R_2\} \\ &\geq R_2^2 \sup_{\bar{\rho}} \{[f_1(\bar{\rho} \pm h, \alpha_0)(1 \mp \sigma/R_2) - \bar{\rho}^{-1}]^2 : \bar{\rho} \geq R_1^{-1}\} \equiv \Delta_\pm^2 \end{aligned}$$

When providing current manipulations, we agree to consider the number α_0 to be a fixed value of the regularization parameter from the range of the function $\alpha_\eta = \alpha_\eta(A_h, u_\sigma, h, \sigma)$. We claim that the quantities $\Delta_\pm$ are finite. Indeed, under the conditions $\bar{\rho} \geq R_1^{-1}$ and $0 \leq h < h_0(R_1) = R_1^{-1}/2$ it turns out that $\bar{\rho} - h > R_1^{-1}/2 > 0$ and, in light of the properties of the function f_1 (see Definition 1), the range of the function $f_1(\bar{\rho} - h, \alpha_0)$ appears to be bounded.

This inequality with regard to $\Delta^2(P, \eta)$ allows us to get the estimate

$$\begin{aligned} \Delta(P, \eta) &\geq \tfrac{1}{2}(\Delta_+ + \Delta_-) \\ &\geq R_2 \tfrac{1}{2}\Big\{|f_1(\rho + h, \alpha_0)(1 - \sigma/R_2) - \rho^{-1}| \\ &\qquad + |f_1(\rho - h, \alpha_0)(1 + \sigma/R_2) - \rho^{-1}|\Big\} \end{aligned} \tag{3}$$

which is valid for $\rho \geq R_1^{-1}$ and $0 \leq h < h_0(R_1)$. Let $\beta \equiv R_1^{-1}$ and

$\gamma \equiv [1+q(h)]R_1^{-1}$. By the original assumption concerning the function $q(h)$ the inequality $\gamma - h > \beta + h$ holds. In view of this, having replaced the variables under the integral sign and having applied the obvious inequality

$$|ax-c|+|bx-d| \geq \frac{|bc-ad|}{\max\{|a|,|b|\}}$$

with $x = f_1(\rho, \alpha_0)$, $a, b = (1 \mp \sigma/R_2) > 0$ and $c, d = (\rho \mp h)^{-1}$ we infer from (3) that

$$\begin{aligned}
\Delta(P,\eta) \;&\geq\; \frac{R_2}{2(\gamma-\beta)} \int\limits_{\beta}^{\gamma} \Big\{ \Big| f_1(\rho+h,\alpha_0)\Big(1-\frac{\sigma}{R_2}\Big) - \frac{1}{\rho}\Big| \\
&\qquad + \Big| f_1(\rho-h,\alpha_0)\Big(1+\frac{\sigma}{R_2}\Big) - \frac{1}{\rho}\Big| \Big\} d\rho \\
&\geq\; R_2 \frac{1}{2(\gamma-\beta)} \int\limits_{\beta+h}^{\gamma-h} \Big\{ \Big| f_1(\rho,\alpha_0)\Big(1-\frac{\sigma}{R_2}\Big) - \frac{1}{\rho-h}\Big| \\
&\qquad + \Big| f_1(\rho,\alpha_0)\Big(1+\frac{\sigma}{R_2}\Big) - \frac{1}{\rho+h}\Big| \Big\} d\rho \\
&\geq\; \frac{R_2(\gamma-\beta-2h)}{2(\gamma-\beta)} \Big[2\Big(h+\frac{\sigma\beta}{R_2}\Big)\frac{1}{\gamma^2}\Big(1+\frac{\sigma}{R_2}\Big)^{-1} \Big] \\
&=\; \Big[1-\frac{2hR_1}{q(h)}\Big](\sigma+R_1R_2h)R_1[1+q(h)]^{-2}\Big(1+\frac{\sigma}{R_2}\Big)^{-1}
\end{aligned}$$

Thus, the lemma is completely proved. □

It is easy to derive from Definition 2 and Lemma 1 one useful corollary.

Corollary 1 *The estimate* $\Delta_0(\eta) \geq \nu(h,\sigma)$ *holds for* $0 \leq h < h_0(R_1)$.

For later comparison, we introduce in accordance with formula (1) and Theorem 1 the accuracy characteristic $\Delta_{\mathrm{mpm}}(\eta)$ of the **m.p.m.** method.

Lemma 2 *As* $h, \sigma \to 0$, *the accuracy characteristic* $\Delta_{\mathrm{mpm}}(\eta)$ *of the* **m.p.m** *method approaches*

$$\Delta_{\mathrm{mpm}}(\eta) = R_1(\sigma + 2hR_1R_2)[1+o(h+\sigma)]$$

In this estimate, the orders of h, σ, R_1, R_2 *and the constant 2 are unimprovable.*

Proof. Observe that Σ contains the data $(\bar{A}, \bar{u})$, what means that $\|\bar{A}^+\|_* \leq R_1$ and $\|\bar{u}\| \leq R_2$. In view of this, $h_0(R_1) = R_1^{-1}/2 \leq \|\bar{A}^+\|_*^{-1}/2 = h_0(\bar{A})$ and the inequality $0 \leq h < h_0(R_1)$ provides the validity of the conditions of Theorem 2.6. Applying Theorem 2.6 yields estimate (2.16) which can be written as

$$\|z_\eta(A_h, u_\sigma, h, \sigma) - \bar{z}\| \leq \|\bar{A}^+\|_* \frac{\sigma + 2h\|\bar{A}^+\|_* \cdot \|\bar{u}\|}{(1-2h\|\bar{A}^+\|_*)^3}$$

where z_η is an approximate solution of the fundamental problem computed using the **m.p.m.** method. The preceding estimate along with (1) implies

$$\Delta_{\mathrm{mpm}}(\eta) \leq R_1[\sigma + 2hR_1R_2/(1-2hR_1)^3] \tag{4}$$

It is clear from Definition 1 that

$$\Delta_{\mathrm{mpm}}(\eta) \geq \|z_\eta(A_h, u_\sigma, h, \sigma) - \bar{z}\|$$

for any exact problem (1.1) with $(\bar{A}, \bar{u}) \in \Sigma$ and any approximate data $(A_h, u_\sigma) \in \Sigma_\eta(\bar{A}, \bar{u})$. To obtain lower estimate for accuracy using the last inequality, we focus our attention on a particular case where the exact data $(\bar{A}, \bar{u})$ are of the form

$$\bar{A} = \mathrm{diag}(R_1^{-1}, 0, \dots, 0) \in \mathfrak{A} \qquad \bar{u} = (R_2, 0, \dots, 0) \in \mathbf{R}^m$$

and the approximate data look as follows:

$$A_h = \mathrm{diag}(R_1^{-1} + h, 0, \dots, 0) \in \mathfrak{A} \qquad u_\sigma = (R_2 - \sigma, 0, \dots, 0) \in \mathbf{R}^m$$

Adopting techniques very similar to those of Section 3 one can construct on the basis of Theorem 3.2 the matrix of the **m.p.m.** method:

$$\widehat{A}_h = \mathrm{diag}[(R_1^{-1} + h)x[\lambda(h)], 0, \dots, 0]$$

The quantity $x[\lambda(h)]$ can be found from the following equation of the form (3.14):

$$\beta(\lambda) = (R_1^{-1} + h)^2[x(\lambda) - 1]^2 = h^2$$

whence it follows that $x[\lambda(h)] = 1 + h/(R_1^{-1} + h)$ and $\widehat{A}_h = \mathrm{diag}(R_1^{-1} + 2h, 0, \dots, 0)$. For the data at hand, we obtain the vectors

$$\bar{z} = \bar{A}^+\bar{u} = (R_1R_2, 0, \dots, 0) \in \mathbf{R}^n$$

$$z_\eta = \widehat{A}_h^+ u_\sigma = \left(\tfrac{R_2-\sigma}{R_1^{-1}+2h}, 0, \dots, 0\right) \in \mathbf{R}^n$$

whose components permits us to derive accurately the estimate

$$\Delta_{\mathrm{mpm.}}(\eta) \geq \|z_\eta - \bar{z}\| = \frac{\sigma R_1 + 2hR_1^2R_2}{1 + 2hR_1} \tag{5}$$

Comparison of (4) with (5) gives the main term of the quantity $\Delta_{\mathrm{mpm}}(\eta)$ as $\eta \to 0$:

$$\Delta_{\mathrm{mpm}}(\eta) \asymp R_1(\sigma + 2hR_1R_2)$$

With this, we finish the proof of the lemma. □

From Lemmas 1–2 and Corollary 1, we establish a profound result.

Theorem 2 *The* **m.p.m.** *method as an algorithm from the class S is of optimal accuracy order*

$$\overline{\lim}_{\eta\to 0} \frac{\Delta_{\mathrm{mpm}}(\eta, R_1, R_2)}{\Delta_0(\eta, R_1, R_2)} \leq 2$$

Proof. From Corollary 1 and (4) it follows that

$$\nu(h,\sigma) \leq \Delta_0(\eta) \leq \Delta_{\text{mpm}}(\eta) \leq R_1 \frac{\sigma + 2hR_1R_2}{(1-2hR_1)^3} \tag{6}$$

Further, the function $q(h)$ from Lemma 1 is taken to be $q(h) = 2\sqrt{h}R_1$. Because of the form of the quantity $\nu(h,\sigma)$ we derived in Lemma 1, relations (6) imply for $0 \leq h < \min\{1, h_0(R_1)\}$ that

$$1 \leq \frac{\Delta_{\text{mpm}}(\eta)}{\Delta_0(\eta)} \leq 2\frac{(1+2\sqrt{h}R_1)^2(1+\sigma/R_2)}{(1-2hR_1)^3(1-\sqrt{h})}$$

Passing on the right to the limit as $h, \sigma \to 0$ we arrive at the assertion of the theorem. □

Observe that relations (6) provide the convergence $\Delta_0(\eta) \to 0$ as $\eta \to 0$.

So, the **m.p.m.** method is of optimal accuracy order with the optimality constant $g = 2$. The reader can compare this method with others by computing similar constants for them. For example, the method from Gilyazov and Morozov (1984) or from Vaĭnikko and Veretennikov (1986) provides

$$g = (7 + 2\sqrt{5})^{1/2} \approx 3.387$$

The **m.p.m.** method possesses one more optimality property known as the **optimal order of proximity of the matrix condition number to the best one**. The condition number of any matrix A is defined to be

$$\nu(A) = \|A^+\|_* \cdot \|A\|$$

The number $\nu(A)$ characterizes stability of the problem of finding a normal pseudosolution of SLAE (1.1) (see, for example, Voevodin (1969), Voevodin and Kuznetsov (1984)). Therefore, from a practical point of view, it is desirable to deal with an approximate matrix A_h whose condition number is smaller than that of any other matrice. In this context, there arises naturally the extremal problem related to a matrix $\breve{A}_h \in \mathfrak{A}_h$ for which

$$\begin{aligned} \nu(\breve{A}_h) &= \inf\{\nu(A) \colon A \in \mathfrak{A}_h\} \\ &= \inf\{\nu(A) \colon A \in \mathfrak{A}, \|A - A_h\| \leq h\} \\ &\equiv \nu_h \end{aligned} \tag{7}$$

The number ν_h is called the **best possible condition number** of problem (1.1) with the approximate matrix A_h.

Theorem 3 *If for a given* $h \geq 0$ *the condition* $\|A_h\| > h$ *holds, then problem* (7) *is solvable.*

Proof. Let $\{A_n\}$ be an arbitrary minimizing sequence for problem (7), that is, $\{A_n\} \subset \mathfrak{A}_h$ and $\nu(A_n) \to \nu_h$ as $n \to \infty$. Then, for $n \geq N_0$, there is a number $\varepsilon(N_0) > 0$ such that

$$\nu(A_n) = \|A_n^+\|_* \cdot \|A_n\| \leq \nu_h + \varepsilon(N_0) \equiv M_0 = \text{const} \tag{8}$$

The set $\{A_n\}$ is compact in the space $\mathfrak{A}$ due to its boundedness:

$$\|A_n\| \le \|A_h\| + \|A_n - A_h\| \le \|A_h\| + h$$

Since $\{A_n\} \subset \mathfrak{A}_h$, the set $\overline{\{A_n\}}$ is closed in that space $\mathfrak{A}_h$ (see Section 2). Therefore, there exists a subsequence $\{A_{n_k}\}$ converging to $A_0 \in \mathfrak{A}_h$ as $k \to \infty$ with $A_0 \ne 0$. Provided the condition $A_0 = 0$ holds, we come to the incompatible results

$$h \ge \lim_{k\to\infty} \|A_{n_k} - A_h\| = \|A_0 - A_h\| = \|A_h\| > h$$

The condition $A_0 \ne 0$ implies the estimate $\|A_{n_k}\| \ge q = \text{const} > 0$ for all $n_k \ge N_1$. Therefore, for all $n_k > \max\{N_0, N_1\}$ (8) yields

$$\|A_{n_k}^+\|_* \le M_0/q \equiv M_1$$

In addition,

$$\|A_{n_k}\| \le \|A_h\| + h \equiv M_2$$

According to Lemma 2.1, combination of the preceding estimates gives $A_{n_k}^+ \to A_0^+$ as $k \to \infty$ and the chain of limit relations

$$\begin{aligned}\lim_{k\to\infty} \nu(A_{n_k}) &= \lim_{k\to\infty} \|A_{n_k}^+\|_* \cdot \lim_{k\to\infty} \|A_{n_k}\| \\ &= \|A_0^+\|_* \cdot \|A_0\| = \nu(A_0) = \nu_h\end{aligned}$$

which justifies that A_0 is a solution of problem (7). Thus, the theorem is completely proved. □

A matrix $\breve{A}_h$ solving problem (7) is said to be **optimal with respect to the condition number on the set** $\mathfrak{A}_h$. It is worth noting here that the set $\mathfrak{A}_h$ contains both elements $\bar{A}$ and A_h. The condition of Theorem 3 is satisfied for $0 \le h < \|\bar{A}\|/2$.

To establish properties of the matrix $\breve{A}_h$ we need an auxiliary lemma.

Lemma 3 *Let p and q be any fixed positive numbers and*

$$\bar{h} \equiv \min\{1, p^{-1}\}q^{-1}\|\bar{A}^+\|_*^{-1}$$

If a matrix A admits the estimates $\|A - \bar{A}\| \le qh$ and $\|A^+\|_* \le p\|\bar{A}^+\|_*$ for $0 \le h < \bar{h}$, then $RgA = Rg\bar{A}$.

The proof is similar to that of Theorem 2.4.

Theorem 4 *The convergence $\breve{A}_h^+ \to \bar{A}^+$ takes place as $h \to 0$. If*

$$\breve{z}_\eta \equiv \breve{A}_h^+ u_\sigma$$

then $\breve{z}_\eta \to \bar{z}$ as $\eta \to 0$.

Proof. Let a number h be such that $0 \le h < \|\bar{A}\|/4$. Then, by Theorem 3, a solution of problem (7) exists and satisfies the estimates

$$\|\breve{A}_h - \bar{A}\| \le \|\breve{A}_h - A_h\| + \|A_h - \bar{A}\| \le 2h$$

$$\|\breve{A}_h^+\|_* = \nu(\breve{A}_h)\|\breve{A}_h\|^{-1} \le \nu(\bar{A})(\|\bar{A}\| - 2h)^{-1} \le 2\|\bar{A}^+\|_*$$

Under the conditions

$$0 \le h < \hat{h} \equiv \min\{\|\bar{A}\|/4, \|\bar{A}^{+}\|_{*}^{-1}/4\}$$

we draw the conclusion by Lemma 3 that the matrices $\check{A}_h$ and $\bar{A}$ both have the same rank. With this, Lemma 2.3 implies the inequality

$$\|\check{A}_h^{+} - \bar{A}^{+}\|_{*} \le \frac{\|\check{A}_h - \bar{A}\| \cdot \|\bar{A}^{+}\|_{*}^{2}}{(1 - \|\check{A}_h - \bar{A}\| \cdot \|\bar{A}^{+}\|_{*})^{3}} \le \frac{2h\|\bar{A}^{+}\|_{*}^{2}}{(1 - 2h\|\bar{A}^{+}\|_{*})^{3}}$$

which provides the required convergence $\check{A}_h^{+} \to \bar{A}^{+}$ as $h \to 0$. By applying an estimate of the form (2.6) to the quantity $\|\check{z}_\eta - \bar{z}\|$ we establish $\check{z}_\eta \to \bar{z}$ as $\eta \to 0$. □

Theorem 4 substantiates stability of the algorithm for approximate calculation of a normal pseudosolution of SLAE (1.1) obtained by means of approximations $\check{z}_\eta$. Moreover, it follows from Theorem 4 that $\nu(\check{A}_h) \to \nu(\bar{A})$ as $h \to 0$.

Although the matrix $\check{A}_h$ possesses an important property of the condition number minimality, the numerical realization of the procedure of constructing such matrices involves some difficulties. To overcome them, it makes sense to apply algorithms of stable calculation of $\bar{z}$ by means of **almost optimal matrices in condition number**. The **m.p.m.** method falls in the category of such algorithms.

Definition 4 *A matrix $A(h) \in \mathfrak{A}_h$ has the* **optimal order of proximity to the best condition number** *if*

$$\nu[A(h)] - \nu(\check{A}_h) \le l h \|\bar{A}^{+}\|_{*}$$

and the constant $l \ge 1$ does not depend on h and $\bar{A}$ both.

Theorem 5 *Any matrix of the* **m.p.m.** *method has the optimal order of proximity to the best condition number.*

Proof. We begin by considering a matrix $\widetilde{A}_h$ of the **m.p.m.** method (see Section 2) and an arbitrary solution $\check{A}_h$ of problem (7). It is straightforward to verify the following estimates

$$\|\widetilde{A}_h^{+}\|_{*} \le \|\bar{A}^{+}\|_{*}$$
$$\|\widetilde{A}_h^{+}\|_{*} \le \|\check{A}_h^{+}\|_{*}$$
$$\|\check{A}_h\| \ge \|\widetilde{A}_h\| - \|\check{A}_h - \widetilde{A}_h\| \ge \|\widetilde{A}_h\| - 2h$$

which serve to motivate the inequality

$$\nu(\check{A}_h) = \|\check{A}_h^{+}\|_{*} \cdot \|\check{A}_h\| \ge \|\widetilde{A}_h^{+}\|_{*}(\|\widetilde{A}_h\| - 2h) \ge \nu(\widetilde{A}_h) - 2h\|\bar{A}^{+}\|_{*}$$

From here, we obtain the relation

$$\nu(\widetilde{A}_h) - \nu(\check{A}_h) \le 2h\|\bar{A}^{+}\|_{*}$$

thereby completing the proof of the theorem. □

6.6 Numerical realization of the m.p.m. method

We confine ourselves to a special case when $\mathfrak{A}_0 = \mathfrak{A}$ and the matrix of the **m.p.m.** method can be obtained in the framework of Section 3. The **algorithm based on the m.p.m. method** incorporates several stages (Leonov (1987 a, 1991):

(1) determining the singular value decomposition for a given matrix A_h: $A_h = U_h D_h V_h^{\mathrm{T}}$;

(2) finding a solution $\lambda(h)$ to equation (3.14);

(3) constructing the matrix $\widehat{A}_h = U_h \widehat{D}_h V_h^{\mathrm{T}}$ of the **m.p.m.** method, where $\widehat{D}_h$ is specified either in Theorem 3.2 or in framework of Theorem 3.4;

(4) calculating the approximate solution $z_\eta = \widehat{A}_h^+ u_\sigma = V_h \widehat{D}_h^+ U_h^{\mathrm{T}} u_\sigma$ and the minimal pseudoinverse matrix $\widehat{A}_h^+ = V_h \widehat{D}_h^+ U_h^{\mathrm{T}}$ if necessary.

In what follows, we discuss each of them in considerable detail. At present, several methods are widely used for the approximate determination of the singular value decomposition of various matrices having the general form (Forsythe *et al.* (1977), Lawson and Hanson (1974), Voevodin (1977)). The suitable computer programs in various high-level languages have been published and exploited in practice (Forsythe *et al.* (1977), Maindonald (1984), Wilkinson and Reinsch (1971)). These methods produce, generally speaking, an **almost singular value decomposition of A_h with the accuracy $\varkappa$**. By a decomposition of this sort, we mean orthogonal matrices $U_{h\varkappa} \in \mathfrak{A}_m$, $V_{h\varkappa} \in \mathfrak{A}_n$ and a diagonal matrix $D_{h\varkappa} = \mathrm{diag}(\rho_1^{h\varkappa}, \dots, \rho_M^{h\varkappa}) \in \mathfrak{A}$ with $\rho_1^{h\varkappa} \geq \dots \geq \rho_M^{h\varkappa} \geq 0$ such that

$$\|A_h - U_{h\varkappa} D_{h\varkappa} V_{h\varkappa}^{\mathrm{T}}\| \leq \varkappa \tag{1}$$

Usually, the quantity $\varkappa$ can be made sufficiently small ($\varkappa \ll h$) to provide a possibility of inserting an almost singular value decomposition in place of the exact one for the algorithm in question. Therefore, we may assume that, instead of the initial approximate matrix A_h, we have at our disposal another matrix $A_{h\varkappa} = U_{h\varkappa} D_{h\varkappa} V_{h\varkappa}^{\mathrm{T}}$ for which the condition of approximation

$$\|\bar{A} - A_{h\varkappa}\| \leq h + \varkappa$$

holds. So, instead of the problem (1.1) approximate data (A_h, u_σ, h) in the **m.p.m.** method, we deal with $(A_{h\varkappa}, u_\sigma, h+\varkappa)$, whose use permits us to replace the data U_h, D_h, V_h and h by $U_{h\varkappa}$, $D_{h\varkappa}$, $V_{h\varkappa}$ and $h+\varkappa$, respectively, in the computational formulae of the algorithm developed in Section 3.

Such a framework eliminates some problems. In essence, finding the exact singular value decomposition of the matrix A_h is equivalent to the whole problem on the eigenvalues of the matrices $A_h^{\mathrm{T}} A_h$ and $A_h A_h^{\mathrm{T}}$ (see Voevodin and Kuznetsov (1984)). When seeking eigenvectors and eigenvalues, we must carry out two tasks concerning accuracy and stability. The algorithms for constructing an almost singular value decomposition allow one to avoid these problems, since it suffices to handle a suitably chosen type of the matrices $U_{h\varkappa}$, $D_{h\varkappa}$ and $V_{h\varkappa}$ satisfying the above conditions. It is concerned, in particular, with condition (1). Then, generally speaking, the column-vectors of the matrices $U_{h\varkappa}$ and $V_{h\varkappa}$ are no longer eigenvectors of the matrices $A_h^{\mathrm{T}} A_h$ and $A_h A_h^{\mathrm{T}}$. However, it does not matter in applications of the **m.p.m.** method.

The second stage of the algorithm in question is connected with solving equation (3.14) with a monotone function

$$\beta(\lambda) = \sum_{k=1}^{M} [\rho_k(\lambda) - \rho_k^{h\varkappa}]^2 = (h + \varkappa)^2 \tag{2}$$

By Lemma 3.2, the left-hand side of equation (2) can be recast as

$$\beta(\lambda) = \sum_{k \in K_1(\lambda)} (\rho_k^{h\varkappa})^2 [x_k(\lambda) - 1]^2 + \sum_{k \in K \setminus K_1(\lambda)} (\rho_k^{h\varkappa})^2 \qquad \lambda > 0 \tag{3}$$

The sets of indices K and $K_1(\lambda)$ are defined by

$$K = \{k\} = \{1, \ldots, M\}$$
$$K_1 \equiv K_1(\lambda) \equiv K_1(\lambda, A_{h\varkappa}) = \{k \in K \colon \lambda(\rho_k^{h\varkappa})^{-4} \leq 27/16\}$$

The quantity $x_k(\lambda)$ represents a positive root to the equation

$$x^4 - x^3 = \lambda(\rho_k^{h\varkappa})^{-4} \tag{4}$$

on the half-open interval $(1, 3/2]$.

The determination of a solution $\lambda(h + \varkappa)$ to equation (2) depends essentially on the way of computing the function $\beta(\lambda)$. The calculation of $\beta(\lambda)$ necessitates imposing, for a given λ, the set of indices $K_1(\lambda)$ in terms of singular values $\rho_k^{h\varkappa}$ of the matrix $A_{h\varkappa}$, seeking a solution $x_k(\lambda)$ to equation (4) for $k \in K_1(\lambda)$ and finding $\beta(\lambda)$ by formula (3). After analyzing the values of the function $\beta(\lambda)$ at its discontinuity points $\lambda_k = 27(\rho_k^{h\varkappa})^4/16$ ($k \in K$) (see Lemma 3.3) and comparing these values with $(h + \varkappa)^2$ one can decide whether a solution to equation (2) is ordinary or generalized.

The search of the generalized solution coinciding with one of the discontinuity points is connected with a finite number of computations of values of the function $\beta(\lambda)$. The procedure of calculating $\beta(\lambda)$ itself requires a relatively small number of operations enabling to apply, for example, the dichotomy method for finding an ordinary solution to equation (2).

The realization of the third stage of the algorithm depends on the sort of a solution $\lambda(h+\varkappa)$ to equation (2), ordinary or generalized. When $\lambda(h+\varkappa)$ is an ordinary solution, the matrix $\widehat{D}_{h\varkappa}$ containing the singular values of the matrix of the **m.p.m.** method takes according to Theorem 3.2 the form

$$\widehat{D}_{h\varkappa} = \operatorname{diag}\{\rho_1[\lambda(h + \varkappa)], \ldots, \rho_M[\lambda(h + \varkappa)]\} \equiv \operatorname{diag}(\rho_1^*, \ldots, \rho_M^*) \in \mathfrak{A}$$

where the functions $\rho_k(\lambda)$ $(k \in K)$ are given by formulae (3.9), according to which we have

$$\rho_k^* = \begin{cases} \rho_k^{h\varkappa} x_k[\lambda(h+\varkappa)] & k \in K_1[\lambda(h+\varkappa)] \\ 0 & k \in K \backslash K_1[\lambda(h+\varkappa)] \end{cases}$$

When $\lambda(h+\varkappa)$ is a generalized solution to equation (2), that is, belongs to discontinuity points of the function $\beta(\lambda)$, the approach justified in Theorem 3.4 proves to be useful. Under the convention

$$\lambda(h+\varkappa) = \lambda_{t(h+\varkappa)+1}$$

(s.f. (3.16)), we may put

$$\widehat{D}_{h\varkappa} = \operatorname{diag}(\rho_1^*, \dots, \rho_M^*)$$

where the numbers ρ_k^* $(k \in K)$ are specified as in Section 3:

$$\rho_k^* = \begin{cases} \rho_k^{h\varkappa} x_k[\lambda_{t(h+\varkappa)+1}] & k = 1, \dots, t(h+\varkappa) \\ 0 & k = t(h+\varkappa)+1, \dots, M \end{cases}$$

By Theorem 3.3, the equality $t(h+\varkappa) = \operatorname{Rg}\bar{A}$ is valid as long as $0 < h+\varkappa < h_1(\bar{A})$.

In conclusion, it should be noted that the pseudoinverse

$$\widehat{D}_{h\varkappa}^+ = \operatorname{diag}[\theta(\rho_1^*), \dots, \theta(\rho_M^*)] \in \mathfrak{A}^*$$

is aimed to calculate the approximate solution $z_{h\varkappa} = V_{h\varkappa}\widehat{D}_{h\varkappa}^+ U_{h\varkappa}^{\mathrm{T}} u_\sigma$.

By virtue of Theorems 2.7 and 3.4, one can establish that $z_{h\varkappa} \to \bar{z}$ as $\eta, \varkappa \to 0$.

It is worth emphasizing here that, when handling a generalized solution $\lambda(h+\varkappa)$ to equation (2), one can produce the matrix of the **m.p.m.** method on the basis of Theorem 3.5 (and Lemma 3.4 and Theorem 3.1). Here the rank of the matrix $\bar{A}$ can be determined by using the stable procedure described in Theorem 3.3.

The total number N of arithmetic operations for finding an approximate solution $z_{h\varkappa}$ in the case of a square $(n \times n)$-matrix A_h can be evaluated as follows:

$$N \asymp 8n^3/3 + 4n^2 + (C_1 l_1 + C_2 l_2)n \quad C_1 = \text{const} > 0 \quad C_2 = \text{const} > 0$$

The first in the last estimate is caused by factorizing the initial matrix as a product of orthogonal and two-diagonal ones in constructing an almost singular value decomposition (see, for example, Forsythe *et al.* (1977)). The second summand is connected with a final computation of the approximation $z_{h\varkappa}$; l_1 iterations are needed to reduce any matrix to a diagonal form in the singular value decomposition (Forsythe *et al.* (1977)); l_2 is the total number of iterations when finding $\lambda(h+\varkappa)$. Quite often, practical computations involve $l_1, l_2 \le C_3 n$ $(C_3 = \text{const} > 0)$. In this case $N \asymp 8n^3/3$. The

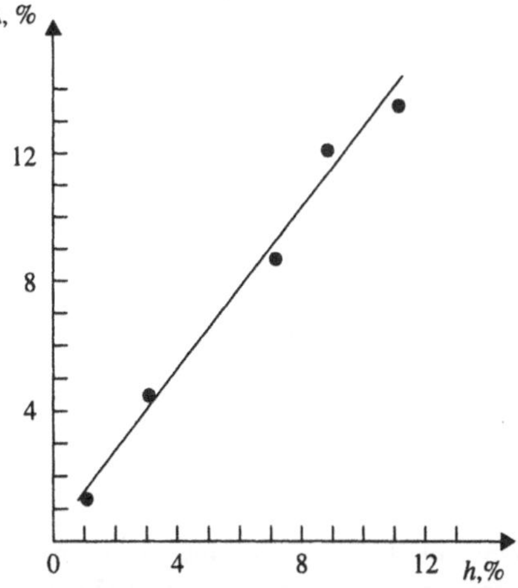

Figure 6.5 The accuracy of recovering the normal pseudosolution of the model system by the m.p.m. method.

total numbers of necessary operations for other well-known stable methods for solving SLAEs are of the same order (see, for example, Section 1 and Gilyazov and Morozov (1984), Tikhonov (1965 e) and Voevodin (1977)).

We now present the results of some model computations by the **m.p.m.** method. The data (points) for the accuracy $\Delta(h)$ of recovering a normal pseudosolution of the model system by the **m.p.m.** principle depending on the matrix perturbation level h are depicted in Fig. 6.5.

It is necessary to clarify several accompanying conditions. The perturbations in matrix are artificially produced by an actual generator of uniformly distributed pseudorandom numbers. The exact matrix $[\bar{a}_{ij}]_{m\times n}$ is of the form $\bar{a}_{ij} = 1$ $(i = 1, \dots, 16, j = 1, \dots, 15)$ and the exact right-hand side $[\bar{u}_i]$ is taken to be $u_1 = 16$, $u_i = 15$, $i = 2, \dots, 15$, and $u_{16} = 14$. We consider it to be unperturbed. The exact normal pseudosolution $[\bar{z}_j]$ of the problem at hand is equal to $\bar{z}_j = 1$. Here, the perturbation level h is evaluated as follows:

$$\begin{aligned} h &= (mn)^{1/2} \max\{|a_{ij}^h - \bar{a}_{ij}| : i = 1, \dots, 16, j = 1, \dots, 15\} \\ &= \sqrt{240} \max\{|a_{ij}^h - \bar{a}_{ij}|\} \end{aligned}$$

where a_{ij}^h are the elements of the perturbed matrix. The quantity h is given as a percentage of $\|[\bar{a}_{ij}]\| = \sqrt{240}$. The accuracy estimation of reconstructing

$$\Delta(h) = \max\{|z_j^h - \bar{z}_j| : j = 1, \dots, 15\}$$

is based on the approximate solution $[z_j^h] = \widehat{A}_h^+ \bar{u}$ obtained by means of a computer program for h which interests us. It turns out that the quantity

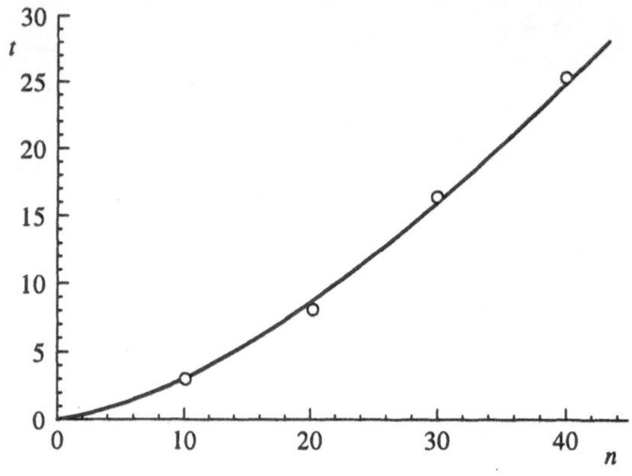

Figure 6.6 Dependence of the time for computing an approximate solution by means of the m.p.m. method on the dimension of the problem.

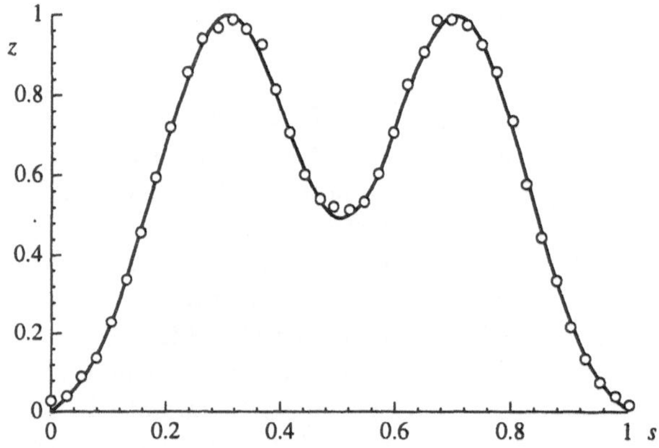

Figure 6.7 The exact solution (—) and approximate solution (∘ ∘ ∘) of the modeling problem obtained by the m.p.m. principle.

$\varkappa$ from (1) is negligibly small in comparison with available values of h. The quantity $\Delta(h)$ in Fig. 6.5 is given as a percentage of $\max\{|\bar{z}_j| : j = 1, \ldots, 15\} = 1$.

Fig. 6.6 shows how the computation time t of approximate solution obtained by the **m.p.m.** principle depends on the dimension n of the problem if h is held fixed. The dependence occurs while solving the model problem $[\bar{a}_{ij}] = [1]$, $[\bar{u}_i] = n$, $i, j = 1, 2, \ldots, n$, with the perturbation level $h = 1\%$, which is calculated in the same manner as before. The dependence involves the time of calculating the minimal pseudoinverse.

Fig. 6.7 illustrates the results of solving a grid analogue of the well-known modeling problem (see Tikhonov *et al.* (1983a, 1995)) with the matrix $[a_{ij}]$

$(i = 1, 2, \dots, m, j = 1, 2 \dots, n)$: $a_{ij} = [1 + 100(x_i - s_j)^2]^{-1}$, $x_i = -2 + 4(i - 1)/(m - 1)$, $s_j = (j - 1)/(n - 1)$; $m = n = 40$. The exact model solution $\bar{z}$ is given by an equality from Tikhonov *et al.* (1983a, 1995)). The right-hand side is found from the matrix and the solution $\bar{z}$ in accordance with (1.1). The matrix is perturbed by random numbers uniformly distributed on the segment $[-0.01, 0.01]$. The estimate $h \leq 0.16$ corresponding to this situation was used in the **m.p.m.** principle. The level of the right-hand side perturbations is equal to $\sigma = 0.013$ (Tikhonov *et al.* (1983a, 1995))). The continuous curve indicates the exact solution.

Let us stress, in conclusion, that the **m.p.m.** method provides an effective tool for solving a large number of SLAEs of the form (1.1) with one and the same approximate matrix A_h and different right-hand sides u_σ. The readers can encounter such situations in processing large series of available experimental data.

CHAPTER 7

Numerical solution of nonlinear ill-posed problems

In this chapter, we share our practical experience of the design in high-quality software of algorithms, whose theoretical background was well-developed in Chapters 2–4 in the process of solving concrete applied problems, with those who wish to use exploratory devices towards a better understanding of real world processes. Because of the enormous range and variety of ill-posed problems dealt with by mathematical physics, it seems expedient to focus the reader's attention only on some of them. The reader can find other examples in the monographs by Bakushinskiĭ and Goncharskiĭ (1989, 1994), Goncharskiĭ *et al.* (1978, 1985), Tikhonov and Arsenin (1977) and the original papers by Kasparov and Leonov (1975), Krekoten' *et al.* (1986), Leonov and Suleĭmanova (1985), etc. In this chapter several modifications of the generalized discrepancy method and the generalized discrepancy principle are offered. In recent years, they have had considerable impact on complex numerical modeling problems.

7.1 Inverse problems in vibrational spectroscopy

Consider inverse problems of vibrational spectroscopy related to the determination of the molecular force field parameters on the basis of experimental data (mainly obtained from an analysis of the molecular vibrational spectra) (see Koptev and Pentin (1977)).

These problems lead to an operator equation of the form

$$Az = u \tag{1}$$

where $z \in Z$ is the set of properties of an object to be determined, $u \in U$ is the set of available experimental data, Z and U are certain spaces and A is an operator specified by making a substantiated choice of the mathematical model of a phenomenon.

In line with the approach of Kochikov, Kuramshina and Yagola (1987) the emphasis in this section is on further development of high-performance regularizing algorithms, whose use permits us to solve the problems posed above.

7.1.1 *Basic equations. The statement of the problem*

The concept of the force field of a molecule emerged from its comprehensive and up-to-date treatment as a quantum-mechanical system consisting of point nuclei and electrons, where the masses of the nuclei and electrons are distinctly different. In this view, it is reasonable to introduce a new small parameter ξ equal to the ratio of the electron mass to the sum of the nuclear masses and to appeal to the adiabatic perturbations theory based on the expansion of all terms of Schrödinger's equation for the molecule in powers of $\xi^{1/4}$. For a rigorous proof of this fact, we refer to Braun and Kiselev (1983). The methodology of second-order equations of perturbation theory guides the choice of the motion of the nuclei in an effective force field $\mathcal{U}(q_1, \dots, q_n)$ produced by the electron subsystem of the molecule (here $q_1, \dots, q_n$ are relative coordinates of the nuclei).

The force field plays an important role in establishing the properties of the molecule. In particular, the equilibrium geometric configuration of the nuclei of the atomes of the molecule (if any) $q_0 = \{q_1^0, \dots, q_n^0\}$ satisfies the relation

$$\left.\left(\frac{\partial \mathcal{U}}{\partial q_i}\right)\right|_{q^0} = 0 \qquad i = 1, \dots, n$$

while the properties connected with small vibrations of the molecule are determined by the matrix of the force constants F:

$$F_{ij} = \left.\left(\frac{\partial^2 \mathcal{U}}{\partial q_i \partial q_j}\right)\right|_{q^0} = 0 \qquad i, j = 1, \dots, n$$

In what follows, we will use experimental data of the vibrational motion of a molecule enabling us to set up the problem of finding the matrix F for a specified equilibrium configuration q^0. The various questions of experimental implementation of the latter are discussed, for instance, in Ioffe *et al.* (1984).

The total number of the generalized coordinates describing the configuration of N nuclei is at least $n \geq 3N - 6$ ($3N - 5$ for a linear molecule). Some coordinates among $q_1, \dots, q_n$ become dependent in the situation when $n > 3N - 6$. Their introduction is justified by the convenience of using the symmetry of the molecule and the interpretation of the results obtained.

The frequencies of the vibrational spectra constitute the main experimental data concerning the molecular vibrations. They are related with the matrix of the force constants F by the eigenvalue equation

$$GFL = L\Lambda \tag{2}$$

where Λ is the diagonal matrix consisting of the squares of frequencies of normal molecular vibrations, that is, $\Lambda = \mathrm{diag}(\omega_1^2, \dots, \omega_n^2)$, G is the matrix of kinetic energy in the momentum representation depending only on the nuclear masses and their equilibrium configuration which is supposed to be known (possibly with a certain error) and L is the matrix of associated eigenvectors.

Within the approximation, the force field of the molecule does not depend on the nuclear masses. The traditional way of treating this case is to consider instead of (2) the following system

$$G_i F L_i = L_i \Lambda_i \qquad i = 1, 2, \ldots, m \tag{3}$$

related to the spectra of m isotopic varieties of the molecule (the subscript i indicates isotopic varieties).

Coriolis' constants characterizing the oscillation-rotation interaction in the molecule contain the important information on the molecular force field. Coriolis' constants are determined by splitting the degenerate oscillation levels and are related to the matrix F in terms of the problem (2) eigenvectors:

$$\zeta = M^{-2} L^* \mathcal{A} \mathfrak{M} \mathcal{A}^* L \tag{4}$$

where ζ is a matrix the elements of which are vectors identical with Coriolis' constants, $\mathfrak{M}$ is a diagonal matrix consisting of the nuclear masses, M is the molecular mass and $\mathcal{A}$ is a matrix giving a precise relationship between the Cartesian displacements of the atoms Δr and the components of q:

$$\Delta r_i = \sum_{k=1}^{n} \mathcal{A}_{ik} q_k, \qquad i = 1, \ldots, N$$

The product of the vector elements in equation (4) is understood in the sense of the vector product. The matrix $\mathcal{A}$ can be recovered from the equilibrium geometry of the molecule. The asterisk symbol stands, as usual, for Hermitian conjugation.

The methods of gas-phase electron diffraction permit us to determine the mean square amplitudes of vibrations of the internuclear distances, of which the squares are the diagonal elements of the matrix

$$\mathcal{D} = H L \Delta L^* H^* \tag{5}$$

where Δ is the diagonal matrix with entries

$$\Delta_i = \frac{\hbar}{2\omega_i} \mathrm{cth} \frac{\hbar \omega_i}{kT} \qquad i = 1, \ldots n$$

Here $\hbar$ is Planck's constant, k is Boltzman's constant, T is the temperature, ω_i is the ith frequency of the vibrational spectrum, H is the matrix with entries

$$H_{mk} = \|R_i^0 - R_j^0\|^{-1} (R_i^0 - R_j^0, \mathcal{A}_{ik} - \mathcal{A}_{jk})$$

where i and j are the indices of the atoms corresponding to the mth interatomic distance and R_i^0 is the radius vector of the ith nucleus in the equilibrium configuration. The matrix H depends only on the arrangement of atoms and the atomic masses.

In just the same way, the dependences of other measured values (in particular, the centrifugal distortion constants) on the matrix F can be written in terms of the eigenfrequencies ω and the matrix L being functions of the elements of F.

We deal with combination of equations (2)–(5) or of some of them depending on available experimental data as a single operator equation with an operator A establishing a correspondence between the real symmetric matrix F and the system of the problem (2) eigenvalues, Coriolis' constants ζ, the amplitudes $\mathcal{D}$, etc.

The set of such data may be viewed as a vector of a finite-dimensional space U. We introduce also the vector z of a finite-dimensional space Z containing either elements of the matrix F or quantities by means of which this matrix can always be parametrized. So, we arrive at the statement of the inverse problem in the framework of Chapter 2.

PROBLEM I. Given equation (1) with $z \in D \subseteq Z, u \in U$, where Z and U are finite-dimensional spaces, D is a closed set of *a priori* constraints of the problem and A is a nonlinear continuous operator on D. It is required to find an approximate solution to equation (1) when instead of A and u we are given their approximations A_h and u_σ with $\|u - u_\sigma\| \leq \sigma$ and $\|Az - A_h z\| \leq \varphi[h, z]$ for all $z \in D$. Here $\varphi[h, z]$ is a known continuous functional tending to 0 as $h \to 0$ uniformly for all $z \in D \cap \bar{S}(0, R)$, where $\bar{S}(0, R)$ is a closed sphere of radius R with center at $z = 0$.

The error in specifying the operator A involves an error in determining the equilibrium configuration of the molecule whose parameters can be found by experiment. A reasonable form of the functional φ will be specified later.

We must emphasize that Problem I, in general, satisfies none of the following conditions which are related to its ill-posed character.

Solvability. It is easily seen that, for example, system (3) is compatible only if

$$\frac{\det G_i}{\det \Lambda_i} = \text{const} \qquad i = 1, \dots, m$$

where m is the total number of isotopomers. This condition may be violated both due to errors in measurements of the frequencies Λ_i and the inexact specification of the geometry G_i or due to the anharmonicity of vibrations ignored by the operator of Problem I when handling experimental data.

The uniqueness of the solution of the problem. In dealing with a single spectrum, we may reduce Problem I to the inverse eigenvalue problem (2). Whence it follows that if G is nonsingular, then any matrix F which can be represented as

$$F = G^{1/2} C \Lambda C^* G^{-1/2} \tag{6}$$

with an arbitrary orthogonal matrix C can be adopted as a solution of Problem I.

Stability of the solution with respect to perturbations of A and u. An example illustrating such instabilities can be easily constructed for a system of the form (3). This fact is well-known to spectroscopists.

But we are somewhat uncertain as to what could be declared to be a solution of Problem I. To facilitate understanding, our next step is the problem of finding a normal pseudosolution of Problem I.

PROBLEM II. It is required to obtain

$$\operatorname{Arg} \inf\{\|z - z_0\| : z \in D, \|Az - u\| = \mu\}$$

where $\mu = \inf\{\|Az - u\| : z \in D\}$.

The element $z_0 \in D$ has to be specified from *a priori* arguments by using both the approximate quantum-mechanical calculations and other ideas (for example, the transferability of the force constants to similar fragments of molecules). It is obvious that, in the case when a solution of Problem I exists and is unique, its pseudosolution is identical with the solution itself. In this context, it is of interest to study one more problem.

PROBLEM III. Given equation (1), we wish to construct from the available data $\{A_h, u_\sigma, h, \sigma\}$ approximations $z_\delta \in D$ to a solution $\bar{z}$ of Problem II for which

$$\lim_{\delta \to 0} \|z_\delta - \bar{z}\| = 0 \qquad \delta \equiv (h, \sigma)$$

That is to say, the algorithm for obtaining z_δ should be regularizing.

7.1.2 The inverse vibrational problem

We now consider the simplest statement of Problem I labeled PROBLEM I' assuming that the vibrational spectrum of a single molecule is known. By an operator A entering equation (1) we mean a rule associating with each vector $z \in \mathbf{R}^{n(n+1)/2}$ composed of the elements of a symmetric $n \times n$-matrix F the ordered set of the eigenvalues of the matrix GF. We regard the ordered set of the squares of frequencies of molecular vibrations as the right-hand side $u \in \mathbf{R}^n$.

PROBLEM II'. It is required to find a normal solution

$$\operatorname{Arg} \inf\{\|z - z_0\| : z \in Z, Az = u\}$$

provided that Problem I' is solvable and, in addition, its solution is not unique (except the case $n = 1$).

Since the operator A involved in equation (1) is completely specified by the matrix G, one might reasonably try to evaluate the deviation of an approximately specified operator A_h (corresponding to some G_ξ) by the error in specifying the matrix G. We may assume $\|G - G_\xi\| \leq \xi$ in a suitably chosen matrix norm. In the space $\mathbf{R}^n$ of the right-hand side's u we introduce the Euclidean norm with weights in a natural way. Suppose that instead of the exact value of the right-hand side u we are given an approximation u_σ with $\|u - u_\sigma\| \le \sigma$. The following theorems on stability of Problems I' and II' have been established by Kochikov, Kuramshina and Yagola (1987).

Theorem 1 *Problem I' is stable in the Hausdorff pseudometric with respect to perturbations of the operator and the right-hand side both.*

Proof. Let Z_0 and Z_η be the sets of the Problem I' solutions in the case when the operator A and the right-hand side u are specified exactly and approximately. We are going to show that

$$\lim_{\eta\to 0} P(Z\eta, Z_0) = 0 \qquad\qquad \eta \equiv (\xi, \sigma)$$

where $P(Z_\eta, Z_0)$, the distance between the sets Z_η and Z_0, is equal to

$$P(Z_\eta, Z_0) = \sup_{x\in Z_\eta} \inf_{z\in Z_0} \|x - z\| + \sup_{z\in Z_0} \inf_{x\in Z\eta} \|x - z\|$$

We refer to the well-known representation for the Problem I' solution, namely to formula (6), where $\Lambda = \operatorname{diag}(u_1, \dots, u_n)$ and C is an arbitrary orthogonal matrix. Likewise, $F_\eta = G_\xi^{-1/2} K \Lambda_\sigma K^* G_\xi^{-1/2}$ with a similar K. In any matrix norm, we might have

$$\begin{aligned}
\inf_{F_\eta} \|F - F_\eta\| &= \inf_K \Big\| G^{-1/2} C \Lambda C^* G^{-1/2} \\
&\qquad -G_\xi^{-1/2} K \Lambda_\sigma K^* G_\xi^{-1/2} \Big\| \\
&\le \Big\| G^{-1/2} C \Lambda C^* G^{-1/2} \\
&\qquad -G_\xi^{-1/2} C \Lambda_\sigma C^* G_\xi^{-1/2} \Big\| \\
&\le \|G_\xi^{-1/2}\| \cdot \|G^{-1/2}\| \cdot \|\Lambda - \Lambda_\sigma\| \\
&\qquad + \|G^{-1/2} - G_\xi^{-1/2}\| \\
&\qquad \times \left(\|\Lambda\| \cdot \|G^{-1/2}\| + \|\Lambda_\sigma\| \cdot \|G_\xi^{-1/2}\| \right)
\end{aligned}$$

Taking into account the relation $\|\Lambda_\sigma\| \le \|\Lambda\| + \sigma$ and exploiting the fact that for all ξ from some segment $[0, \xi_0]$ the estimates

$$\|G_\xi^{-1}\| \le \|G^{-1}\| \left(1 - \xi \|G^{-1}\|\right)^{-1} \qquad \|G^{-1/2} - G_\xi^{-1/2}\| \le f(\xi)$$

are valid with $f(\xi)$ being a continuous monotonically increasing function with $f(0) = 0$, we eventually get

$$\begin{aligned}\inf_{F_\eta} \|F - F_\eta\| &\leq f(\xi)\|\Lambda\| \cdot \|G^{-1/2}\| \\ &+ \left[\sigma\|G^{-1/2}\| + f(\xi)(\|\Lambda\| + \sigma)\right]\|G^{-1}\|^{1/2} \\ &\times \left[1 - \xi\|G^{-1}\|\right]^{-1/2}\end{aligned}$$

The norms for z and F as well as for u and Λ being equivalent imply that

$$\sup_{z \in Z_0} \inf_{x \in Z_\eta} \|x - z\| \leq k_1 f(\xi) + k_2\sigma + k_3 f(\xi)\sigma$$

where k_1, k_2 and k_3 are certain constants. The same procedure works for the second summand in the expression for the Hausdorff pseudometric. Thus,

$$P(Z_\eta, Z_0) \leq k_1' f(\xi) + k_2'\sigma + k_3' f(\xi)\sigma \longrightarrow 0 \qquad \xi, \sigma \to 0$$

□

Theorem 2 *Problem II' is stable with respect to perturbations of the right-hand side and the operator both.*

It is worth noting that if a normal solution of Problem I' may be non-unique, it is assumed to exist in the form of β-convergence of approximate solutions to the set of all normal solutions of the exact problem.

In proving, we keep in mind that, for any η, in the η-neighborhood of $\bar{z}$ there exists an element $\tilde{z} \in Z_\eta$ which possesses the normality property with respect to z_0. If this were not the case, we could find in a neighborhood of the solution of the problem with the approximate data $\bar{z}_\eta$ an element of Z_0 for which the distance to z_0 is smaller than that from $\bar{z}$ to z_0. In our reasonings, it is essential that the sets Z_0 and Z_η are closed and bounded in general.

7.1.3 Regularizing algorithms for the problem of finding a normal pseudosolution

Consider Problem I in a Hilbert space (Problem I") in which we wish to construct from a specified set of approximate data $\{A_h, u_\sigma, h, \sigma\}$ approximations z_δ to a solution $\bar{z}$ satisfying the limit relation $z_\delta \to \bar{z}$ as $\delta \equiv (h, \sigma) \to 0$. We must succeed in showing that the methods developed in Chapters 2–4 apply equally well to the numerical solution of that problem.

We offer below some modification of the generalized discrepancy principle with a new definition of the operator error:

$$\|Az - A_h z\| \leq h\|A_h z\| \equiv \varphi[h, z] \tag{7}$$

Thereby we impose the relative error of Az and a more convenient estimate for the problem under consideration than a monotone function of the form $\psi(h, \|z - z_0\|)$.

Furthermore, z^α is taken to be an extremal, not necessarily unique on the set D, of the smoothing functional

$$M^\alpha[z] = \|A_h z - u_\sigma\|^2 + \alpha\|z - z_0\|^2 \tag{8}$$

Observe that its existence follows immediately from the results of Chapter 3. We now deal with function

$$\rho(\alpha) = \|A_h z^\alpha - u_\sigma\| - \frac{1}{1-h}\Big[\hat{\mu}_\delta + k(\sigma + h\|u_\sigma\|)\Big] \tag{9}$$

where $k > 1$ is a constant and

$$\hat{\mu}_\delta = \inf\{\|A_h z - u_\sigma\| + \sigma + h\|A_h z\| \colon z \in D\}$$

Provided that the condition

$$\|A_h z_0 - u_\sigma\| > \frac{1}{1-h}\Big[\hat{\mu}_\delta + k(\sigma + h\|u_\sigma\|)\Big] \tag{10}$$

holds, the equation $\rho(\alpha) = 0$ possesses a generalized solution $\alpha_\delta > 0$. In other words, we choose a parameter α_δ such that $\rho(\alpha) > 0$ if $\alpha > \alpha_\delta$ and $\rho(\alpha) < 0$ if $\alpha < \alpha_\delta$. We claim that $\rho(\alpha_\delta) = 0$ when α_δ belongs to the set of the continuity points of the function $\rho(\alpha)$. Indeed, this assertion follows from the monotonicity of $\rho(\alpha)$ and the limit relations as $\alpha \to +0$ and $\alpha \to +\infty$.

In accordance with what has been said we can design the algorithm for constructing approximations to a normal solution of Problem I".

Step 1. If condition (10) fails to hold, we set $z_\delta = z_0$.

Step 2. If condition (10) is satisfied, we look for a generalized solution $\alpha_\delta > 0$ to the equation $\rho(\alpha) = 0$ and then accept $z_\delta = z^{\alpha_\delta}$. If functional (8) possesses more than one extremal, we choose among them the one for which

$$\|A_h z^{\alpha_\delta} - u_\sigma\| \le \frac{1}{1-h}\Big[\hat{\mu}_\delta + k(\sigma + h\|u_\sigma\|)\Big]$$

The reasoning is the same as in Section 3.2 in Chapter 3.

In what follows we will prove the lemma which is an analogue of Theorem 2.3.1.

Lemma 1 *Let the conditions of Problem I" hold and*

$$\hat{\mu}_\delta \equiv \hat{\mu}(A_h, u_\sigma) = \inf\{\|A_h z - u_\sigma\| + \sigma + \varphi[h, z] \colon z \in D\}$$

Then $\hat{\mu}(A_h, u_\sigma) \ge \mu$ *and*

$$\lim_{\delta \to 0} \hat{\mu}_\delta = \mu \tag{11}$$

Proof. With the aid of relations

$$\mu = \inf\{\|Az - u\| \colon z \in D\} \le \|Az - u\| \le \|A_h z - u_\sigma\| + \sigma + \varphi[h, z]$$

which are valid for any $z \in D$, we obtain

$$\mu \leq \inf\{\|A_h z - u_\sigma\| + \sigma + \varphi[h, z] \colon z \in D\} \equiv \hat{\mu}_\delta$$

By the same token,

$$\hat{\mu}_\delta \leq \|A_h \bar{z} - u_\sigma\| + \sigma + \varphi[h, \bar{z}] \leq \mu + 2(\sigma + \varphi[h, \bar{z}])$$

yielding

$$\mu \leq \hat{\mu}_\delta \leq \mu + 2(\sigma + \varphi[h, \bar{z}])$$

With these relations established, we arrive at (11). □

Theorem 3 ((Kochikov, Kuramshina and Yagola (1987))) *The algorithm so constructed is regularizing.*

Proof. Let $\bar{z} = z_0$. Then it is easy to demonstrate that condition (10) fails to hold and $z_\delta = z_0$. But if $\bar{z} \neq z_0$, then $\|Az_0 - u\| = \mu + \varepsilon (\varepsilon > 0)$. Whence it follows that

$$\|A_h z_0 - u_\sigma\| \geq \frac{\mu + \varepsilon - (\sigma + h\|u_\sigma\|)}{1 + h} \tag{12}$$

As $h, \sigma \to 0$, the right-hand side of (12) approaches $\mu + \varepsilon$, while the right-hand side of (10) tends to μ. Just for this reason, condition (10) will be true once started with some h, σ.

Consider a generalized solution $\alpha_\delta > 0$ to the equation $\rho(\alpha) = 0$. Let an extremal z^{α_δ} of functional (8) be nonunique. We choose from the set of extremals an element $z_+^{\alpha_\delta}$ for which

$$\|A_h z_+^{\alpha_\delta} - u_\sigma\| \geq \frac{1}{1-h} [\hat{\mu}_\delta + k(\sigma + h\|u_\delta\|)]$$

The condition $M^{\alpha_\delta}[z^{\alpha_\delta}] \leq M^{\alpha_\delta}[\bar{z}]$ provides the validity of the following relations

$$\begin{aligned} \xi(h, \sigma) &\equiv \frac{\left[\hat{\mu}_\delta + k(\sigma + h\|u_\sigma\|)\right]^2 - (\hat{\mu}_\delta + h\|u_\sigma\| + \sigma)^2}{(1-h)^2 \alpha_\delta} \\ &\leq \|\bar{z} - z_0\|^2 - \|z_+^{\alpha_\delta} - z_0\|^2 \end{aligned}$$

and $\|z_+^{\alpha_\delta} - z_0\| \leq \|\bar{z} - z_0\|$. It is possible to verify that

$$\lim_{h, \sigma \to 0} \xi(h, \sigma) = 0$$

Choosing an extremal $z_-^{\alpha_\delta}$ for which

$$\|A_h z_-^{\alpha_\delta} - u_\sigma\| \leq \frac{1}{1-h} \left[\hat{\mu}_\delta + k(\sigma + h\|u_\sigma\|\right]$$

and making use of the relations $\xi(h, \sigma) \to 0$ as $h, \sigma \to 0$ and $M^{\alpha_\delta}[z_-^{\alpha_\delta}] \leq M^{\alpha_\delta}[\bar{z}]$ we eventually get

$$\overline{\lim}_{\delta \to 0} \|z_-^{\alpha_\delta} - z_0\| \leq \|\bar{z} - z_0\|$$

Extracting from the set $\{z_-^{\alpha_\delta}\}$ a weakly convergent subsequence, it is not difficult to establish the weak convergence of $z_-^{\alpha_\delta}$ to $\bar{z}$ using the limit relation

$$\lim_{\delta\to 0} \|Az_-^{\alpha_\delta} - u\| = \mu$$

Applying the scheme of compact embeddings or requiring the extra property of the strong continuity for the operators A and A_h, we arrive at the assertion of the theorem. □

For versions of Problem I", in which the estimate for the error of the operator cannot be represented in the form (7), but the conditions of Problem I" are satisfied, the following version of the generalized discrepancy method can be used for infinite-dimensional spaces Z and U.

PROBLEM IV. It is required to find

$$\text{Arg inf}\{\|z - z_0\| : z \in Z_\delta\}$$
$$Z_\delta \equiv \{z \in D : \|A_h z - u_\sigma\| \le \sigma + \varphi[h, z] + \hat{\mu}_\sigma\}$$

where the upper bound for the measure of incompatibility of the exact problem is given by

$$\hat{\mu}_\sigma = \inf\{\|A_h z - u_\sigma\| + \sigma + \varphi[h, z] : z \in D\}$$

Lemma 2 *Problem IV is solvable for any $u_\sigma \in U$ with $\|u - u_\sigma\| \le \sigma$ and for any continuous operator A_h satisfying the relation $\|A_h z - Az\| \le \varphi[h, z]$ for all $z \in D$.*

Proof. As $\bar{z} \in Z_\delta$, the proof of the lemma follows from the nonemptiness of the set Z_δ for any fixed δ because it is bounded and closed too. The set, on which $\|z - z_0\|$ is being minimized, may be made bounded by introducing a new set $Z_\delta \cap \bar{S}(z_0, \|\bar{z} - z_0\|)$. □

Theorem 4 *The algorithm defined by the extremal Problem IV is a regularizing one for Problem I".*

Proof. The proof is similar to that of Theorem 3.5.4 and follows from the continuity of the norm $\|Az_\delta - u\|$ with respect to z_δ on the basis of the relation

$$\|z_\delta - z_0\| \le \|\bar{z} - z_0\|$$

which is valid for all δ. For infinite-dimensional spaces, one should either require that the set D is convex and the operators A and A_h are strongly continuous or use the scheme of compact embeddings. □

The proposed algorithms provide the β-convergence of the approximations to the set of the Problem I" solutions obtained above in the case when its solution is not unique.

7.1.4 The estimate of the operator error

For the operator of the inverse vibrational problem corresponding to equations (2)–(3) it is possible to evaluate the error in the form (7). Simultaneously, the estimate

$$\|Az - A_h z\| \le h_1 \|z\| \tag{13}$$

is true as $h, h_1 \to 0$ provided that the errors in determining the equilibrium configuration of the molecule are being decreased.

Suppose instead of the exactly given matrix G (see Section 1.1) we have at our disposal a matrix G_ξ satisfying the relation $\|G - G_\xi\| \le \xi$ in a suitable matrix norm. Let the matrix G correspond to an exactly specified operator A and the matrix G_ξ correspond to an approximately specified operator A_h.

By the well-known theorem on perturbations of the eigenvalues of an ordinary matrix A (see Lancaster and Tismensky (1985)),

$$|\tilde{\mu}_j - \mu_j| \le \|B\| \max\{\nu(A), \nu(C)\}$$

where $C = A + B$ is an ordinary matrix, μ_j are the eigenvalues of the matrix A, $\tilde{\mu}_j$ are the eigenvalues of the matrix C and $\nu(A)$ is a known scalar function of A. Applying this theorem to $A = GF$ and $C = G_\xi F$ yields

$$|\tilde{\mu}_j - \mu_j| \le \|F\| \cdot \|G - G_\xi\| \max\left\{\varkappa^{1/2}(G), \varkappa^{1/2}(G_\xi)\right\}$$

where $\varkappa(G)$ is the condition number of the matrix G. With the aid of the relation

$$\|Az - A_h z\|^2 = \sum_j |\tilde{\mu}_j - \mu_j|^2 p_j$$

where p_j are some positive weights, we arrive at assertion (13) implying estimate (7) in accordance with formula (6).

7.1.5 Numerical realization of methods

We know from Section 1.2 that Problem II' is stable and can employ for its solving several methods similar to the Monte–Carlo method for which the implementation in high-quality software is simpler and more well-developed. Such an approach is described in detail in Kochikov, Kuramshina *et al.* (1986) and is omitted here.

We now describe widely used numerical methods relating to the problems of Section 1.3. The solution of the problem produced by minimizing the smoothing functional (8) was obtained by a modified method of conjugate gradients in the case of a nonquadratic functional with regard to *a priori* constraints of the form $z_i^* \le z_i \le z_i^{**}$, where z_i^* and z_i^{**} are certain constants.

Each step of the minimization algorithm related to a functional $f(z)$ is implemented with a knowledge of $z^{(k)}$ belonging to the domain of constraints. By means of these values, the remaining parameters are determined in the following way:

$$s^{(k)} = -f'[z^{(k)}] + \frac{s^{(k-1)}\|f'[z^{(k)}]\|^2}{\|f'[z^{(k-1)}]\|^2}$$
$$\lambda_{max} = \min\{(\lambda_i, \mu_j), i \in I_+, j \in I_-\}$$
$$I_+ = \{i\colon s_i^{(k)} > 0\} \quad I_- = \{i\colon s_i^{(k)} < 0\}$$
$$\lambda_i = \frac{-z_i^* + z_i^{(k)}}{s_i^{(k)}} \qquad \mu_j = \frac{z_j^{(k)} - z_j^{**}}{s_j^{(k)}}$$

Here $s^{(0)} = -f'[z^{(0)}]$.

In passing we must solve the problem

$$\text{Arg}\ \min\{f(z^{(k)} + \lambda s^{(k)})\colon \lambda \in [0, \lambda_{max}]\}$$

The idea here is to use the recurrence relation

$$z^{(k+1)} = z^{(k)} + \lambda^* s^{(k)}$$

where $\lambda^* \geq 0$ is an unknown solution which is sought.

When $z^{(k+1)}$ is located on the boundary of the constraints domain, at the next step there is no need to carry out minimization with respect to those components of z for which $z_i^{(k+1)} = z_i^*$ or $z_i^{(k+1)} = z_i^{**}$.

As a result, the dimension of the set, where the functional is being minimized, decreases until the minimum on some subspace of Z is attained. After that, depending on the sign of $f_i'(z)$, the missing component z_i may be involved in the minimization path again.

To provide the decrease of functional $f(z)$ in the direction s it is fairly common to keep $s = -f'(z)$ from time to time. This approach could be useful in the case when the dimension of the set on which minimization is accomplished varies arbitrarily.

One of the possible ways of solving the equation $\rho(\alpha) = 0$ (see (9)) is to decrease α from a certain value $\alpha_0 > 0$ for which $\rho(\alpha_0) > 0$.

To solve the extremal problem IV with nonconvex constraints we employ the linearization method (see Pshenichny (1993)) suitable to the problem

$$\text{Arg}\ \inf\{\|z - z_0\|^2\colon z \in D, f(z) \leq 0\}$$

where $f(z)$ is the same as in the Problem IV statement.

Each step of the algorithm consists of studying the following two problems by relating $z^{(k)} \in D$ to be determined earlier:

$$\text{Arg}\inf_p\{2(z^{(k)} - z_0, p) + \tfrac{1}{2}(p, p)\colon (f'(z^{(k)}), p) + f(z^{(k)}) \leq 0\} \tag{14}$$

and

$$\operatorname{Arg}\inf_{\lambda}\{\|z^{(k)} - z_0 + \lambda p_k\|^2 + N\mathcal{F}(z^{(k)} + \lambda p_k) \colon \lambda \geq 0\} \tag{15}$$

with p_k being a solution of (14) and $\mathcal{F}(z) = \max\{0; f(z)\}$. Here a solution λ^* of problem (15) is needed to calculate the next value by the recurrence formula $z^{(k+1)} = z^{(k)} + \lambda^* p_k$.

It is worth noting here that problem (14) admits an analytical solution. Namely, if $\bar{p} \equiv -2(z^{(k)} - z_0)$ obeys the constraints imposed at the very beginning, then $p_k = \bar{p}$. Otherwise,

$$p_k = \bar{p} - \frac{f(z^{(k)}) + (f'(z^{(k)}), \bar{p})}{\|f'(z^{(k)})\|^2} f'(z^{(k)})$$

In fact, the coefficient at the term $-f'(z^{(k)})$ is the Lagrange multiplier of the problem under consideration; its explicit form enables one to choose properly the parameter N entering the algorithm.

A software package for data processing has been produced on the basis of these algorithms. It incorporates programs for reading information, for auxiliary data preparation, for solving the inverse problem, for solving the direct problem and the dispatcher program. The implementation of the software package is described in more detail in Kochikov and Kuramshina (1985). Here we point out only its main stages. First, it is necessary to provide the input of the user-prepared information about the equilibrium geometry and its symmetry as well as about the available experimental data. Second, one is to calculate the auxiliary matrices $G, \mathcal{A}, H$ (see Section 1.). Problem I'' is solved by minimizing functional (8) for various values of α with the proper choice of the parameter α. As an alternative way, the methodology of the linearization method allows us to solve Problem IV. Finally, the results of the user's interest are printed.

As an example we present some information about calculations of the matrix of force constants for the molecule CH_3SiH_3 and its isotopic modifications CH_3SiD_3, CD_3SiH_3 and CD_3SiD_3. By physical reasoning we imagine that 16 independent parameters of the force field among 50 vanish and the available experimental data consist of 48 frequencies of molecular vibrations (12 frequencies per each modification). As has been noted in Section 7.1.1, this problem appears to be incompatible.

As a result of seeking approximations to a normal pseudosolution of the problem we have obtained the matrix F_δ of force constants reproducing the experimental information within 1–5%. This matrix is the closest one to the *a priori* specified diagonal matrix F_0. The maximal nondiagonal element of F_δ did not exceed $\frac{1}{4}$ and in most cases $\frac{1}{10}$ of the diagonal elements values. A more detailed description of calculations performed with the use of the program packadge is available in the papers by Kochikov, Kuramshina *et al.* (1984) and Kochikov, Yagola *et al.* (1984, 1985).

7.2 Optimization of medicine therapy regimen

The problem of the medicine dosages is one of the most important practical problems arising in clinical trials and life-testing experimentation. This is especially true for medicines possessing a 'narrow effect' when even a small overdosage may result in serious side-effects. On the other hand, there is no positive therapeutic effect when taking a small underdosage. A suitably chosen dosage depends essentially on individual peculiarities of a patient (sex, age, the body's mass, the kind of a pathology and more). Unfortunately, some instructions to medicines disregard these specificities.

Numerous medicines intended for the treatment of acute infections appear to be toxic to a certain extent. For this reason, the approved dosages of such medicines should not only satisfy the conditions of optimal effect, but also guarantee its minimal expenditures. The above considerations are also caused by the fact that quite often a broad and insufficiently substantiated large dosage of brand new medicine may give rise to the growing stability of disease microbes to the medicine which reduces its efficiency.

A careful analysis of a hospital daily routine allows us to give the mathematical statement of the optimization problem of medicine dosages. The materials present the reader with 'real-life' situation. One of the possible versions (see Korotaev *et al.* (1986)) is based on the function $c(t)$ depending on the time t and making the sense of a certain therapeutic medicine concentration in an inflammation seat. Let the time dependence of medicine doses taken by a patient at each moment be well-characterized by a function $u(t)$. We imagine 'a black box' as the simplest model of a human body and write a reasonable relationship between the 'input signal' u and the 'output function' c:

$$c(t) = \int_0^t K(t-\tau)u(\tau)d\tau \qquad 0 \le t \le T \tag{1}$$

A function $K(t)$ being the weight (transfer) function of the organism ('black box') indicates his/her 'response' to the δ-shaped input signal. Here the functions $K(t)$ and $c(t)$ defined for $t \ge 0$ are continuous and nonnegative with $K(0) = 0$.

The class of all functions $u(t)$ taking the form

$$u(t) = \sum_{k=1}^{N} u_k \delta(t - t_k) \tag{2}$$

where N is the total number of doses, t_k is the time of taking the medicine, $0 < t_1 \le \cdots \le t_N \le T_0 < T, T_0 = \text{const}$, and $u_k \ge 0$ is a medicine dose taken by the patient at moment t_k coinciding with a typical value for the problem at hand. From a medical point of view, the specific form (2) of input functions $u(t)$ can be explained by the fact that the time of the usual dissolution of pellets in the alimentary canal is evaluated in seconds, while the time of excreting the medicine from a human body after a single dosage is several hours.

The function $K(t)$ can be recovered by experiment as the reaction of a person to a single unit dosage. In this case the relationship between $c(t)$ and $u(t)$ is of an alternative form

$$c(t) = \sum_{k=1}^{N} K(t - t_k)u_k \tag{3}$$

In many particular cases one can determine on the basis of specific medical reguirements the form of the function $c(t)$ corresponding to a necessary therapeutic effect. In this situation, there arises a possibility of solving the inverse problem, i.e. from the available function $c(t)$ it is necessary to obtain times t_k and doses u_k so that condition (3) holds true. Equation (3) is unsolvable, as a rule, with respect to (t_k, u_k) under the standard constraints imposed on the set of unknowns. Therefore, instead of equation (3) one should consider the extremal problem in which it is required to find times $t_k^*(0 < t_1^* \leq \cdots \leq t_N^* \leq T_0)$ and doses u_k^* satisfying the relation

$$\begin{aligned} \Phi(t_1^*, \dots, t_N^*; u_1^*, \dots, u_N^*) \quad = \quad & \inf\bigl\{\Phi(t_1, \dots, t_N; u_1, \dots, u_N): \\ & 0 < t_1 \leq \cdots \leq t_N \leq T_0; u_k \geq 0\bigr\} \end{aligned} \tag{4}$$

where

$$\Phi(t_1, \dots, t_N; u_1, \dots, u_N) = \left\| \sum_{k=1}^{N} K(t - t_k)u_k - c(t) \right\|^2_{L_2[0,T]}$$

One can show that problem (4) is solvable if the inequality $K(t^*) > 0$ is true at some point $t^* \in (0, T - T_0)$ (see Korotaev *et al.* (1986)).

The usual hospital practice is to fix the times of taking the medicine by a patient. We will denote them by $t_k, k = 1, \dots, N$. Under this agreement, we may set up another extremal problem in which it is necessary to obtain doses u_k^{**} such that

$$\Phi_0(u_1^{**}, \dots, u_N^{**}) = \inf\bigl\{\Phi_0(u_1, \dots, u_N): u_k \geq 0, k = 1, \dots, N\bigr\} \tag{5}$$

where $\Phi_0(u_1, \dots, u_N) \equiv \Phi(t_1, \dots, t_N; u_1, \dots, u_N), t_k = \text{const}$. Problem (5) is also solvable under standard restrictions on the function $K(t)$.

Since problems (4)–(5) happen to be unstable with respect to perturbations of the data (K, c) and nonuniquely solvable, the selection rule for Ω-optimal solutions (see Chapter 2) becomes rather urgent. In the framework of Chapter 2 of the first volume we may attempt a stabilizing functional in the form

$$\Omega(t_1, \dots, t_N; u_1, \dots, u_N) = \sum_{k=1}^{N} \left[p_k(t_k - t_k^0)^2 + q_k(u_k - u_k^0)^2\right] \tag{6}$$

where the values t_k^0 and u_k^0 correspond to some approved regime of taking the medicine. Here we consider the weight coefficients $p_k \geq 0$ and $q_k > 0$

to be fixed and keep $p_k = 0$ for problem (5). Functional (6) accounts the deviation of regime (2) with parameters t_k and u_k from a known in advance regime with parameters t_k^0, u_k^0. By means of Ω-optimal solutions of problems (4)–(5) one can find the regimes of taking the medicine that not only provide the best approximation of the therapeutic concentration occurring in the infection seat in accordance with model (3) to a prescribed function $\bar{c}(t)$, but also give the minimal deviation from the regimes accepted as standard in the usual practice.

On the other hand, if we agree to consider

$$\Omega(u_1, \dots, u_N) = \sum_{k=1}^{N} q_k(u_k)u_k \qquad q_k(u) \geq q_0 = \text{const} > 0 \tag{7}$$

then Ω-optimal solutions of problem (5) depending on the choice of the function $q_k(u)$ permit us to establish the regime of taking the medicine with minimal toxic effects (in this case $q_k(u)$ is to be determined by experiment) or, for instance, with the minimal cost of the expended medicine (if $q_k(u) = q_0 = \text{const}$).

Modern computers allow the implementation of the optimization process of the medicine dosages by solving numerically the ill-posed extremal problems (4)–(5) in a hospital for the treatment of patients with acute inflammatory kidney diseases (pyaelonephritis). Computer programs for constructing Ω-optimal solutions of problems (4)–(5) are based on the **g.p.d.** algorithm developed in Chapter 2 of the first volume for solving extremal problems. The Fletcher–Reeves method can serve here as one possible minimization method (see Polak (1971)). The regularization parameter chosen by the generalized discrepancy principle as a result of solving a number of typical problems is considered to be fixed.

The main condition of the proper treatment of the inflammatory kidney process is creating and maintaining the constant medicine concentration at a prescribed level. For instance, it is known that the treatment of patients with pyaelonephritis is rewarding for $\bar{c}(t) = 150 \pm 50$ mcg/ml (see Korotaev *et al.* (1986)) when taking the medicine 5-NOK. From the function $\bar{c}(t) = 150$ and the experimentally recorded dependence $K(t)$ one can determine by means of the **g.p.d.** algorithm the regime of taking the medicine being optimal in the sense of the minimal toxic effect. Especial attention is being paid to some concrete results.

Fig. 7.1 shows the experimental function $K(t)$ for one of the patients. In Fig. 7.2 the optimized medicine doses are presented; these data were obtained as a result of computer calculations of the problem (5) solution being optimal by some functional Ω of the form (7) (the minimal toxic effect). This regime was successfully realized in practice. The experimental dependence $c(t)$ as a final result of such observations provided by this dosage is shown diagrammalically in Fig. 7.3. It is easily seen that almost all values of $c(t)$ get into the admissible scatter band ± 50 mcg/ml near the

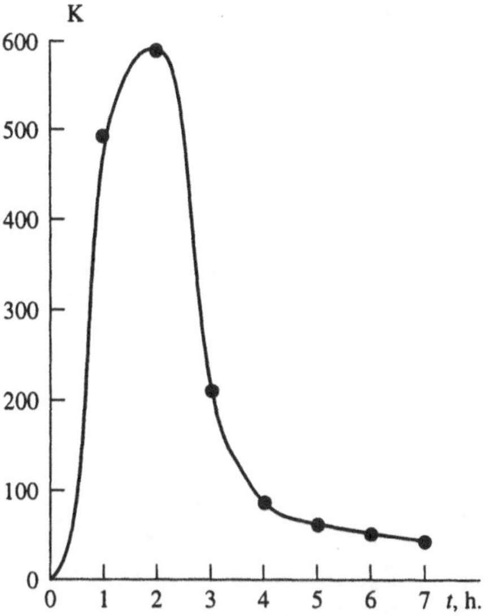

Figure 7.1 Experimental weight function.

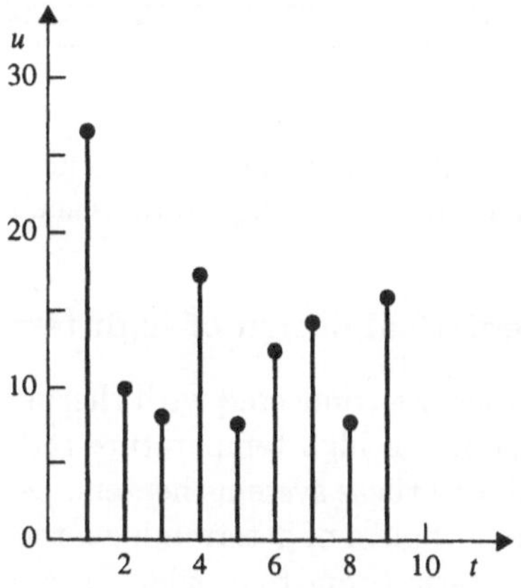

Figure 7.2 The optimized regime of taking the medicine.

value $\bar{c}(t)$. A comparison of the experimental dependence $c(t)$ for the same patient is plotted in Fig. 7.4 in the case when he followed recommendations of the producing firm. Observe that almost at all points the concentrations $c(t)$ exceed in 3–7 times the optimal value $\bar{c}(t)$.

The optimization process of dosing in which the daily dose of the medicine is 3–5 times lower than the prescription of the producing firm permits us to allay the toxic effect. In addition, calculations for the time optimization being used to receive the fixed medicine doses were carried out.

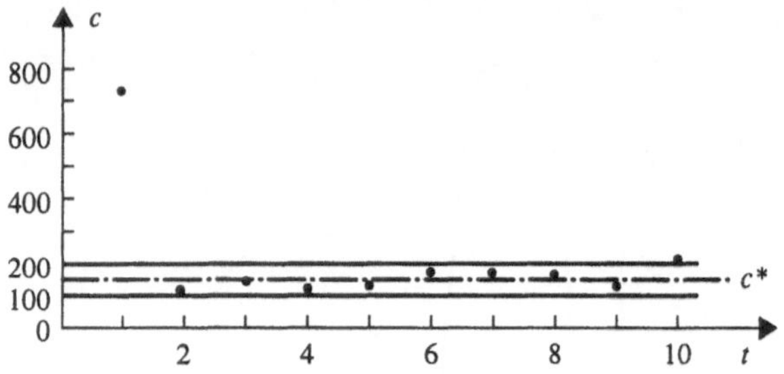

Figure 7.3 The experimental dependence $c(t)$ for the optimized treatment regime.

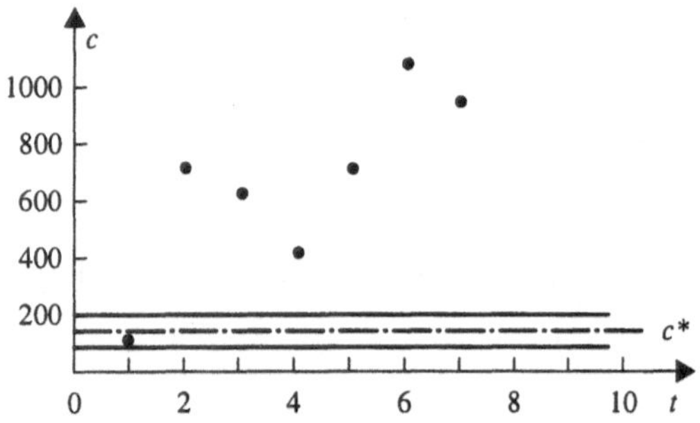

Figure 7.4 The experimental dependence $c(t)$ for the treatment regime recommended by the producing firm and applied to the same patient.

7.3 Optimal mathematical design of high temperature radiators

Many technological processes connected with thermal treatment of objects widely use various systems of high temperature radiators as heat sources. In what follows we will call these systems heaters. Development of concrete heaters is related to the problem of determining a temperature field at their surfaces whose support permits one to provide a necessary temperature field as well as heat flows on an object. Common practice involves mathematical models capable of describing that problem in a certain approximation.

We now deal with surfaces of an object and of a heater, respectively, say F_1 and F_2. Without loss of generality we may assume that a smooth surface F_1 and a plane surface F_2 are not intersecting. The temperature fields $T(M_1)$ and $T(M_2)$ are defined on the surfaces F_1 and F_2, respectively (here M_i is a point on the surface F_i, $i = 1, 2$). In subsequent reasonings the field $T(M_1)$ is supposed to be given, while an unknown field $T(M_2)$ has to be found. In addition, it is assumed that the surfaces F_1 and F_2 radiate and reflect radiation diffusively and selectively (see Rusin and Peletskiĭ (1987)).

On either of these surfaces F_i the dependences of the emittivity $\varepsilon =$

$\varepsilon_i(T,\lambda)$ upon the temperature T and the wavelength of the radiation λ are known in advance. Then the reflectivity $r_i(T,\lambda) = 1 - \varepsilon_i(T,\lambda)$ is also available and according to Kirchhoff's law (Rusin and Peletskiĭ (1987)) the absorptivity is well-defined by $a_i(T,\lambda) = \varepsilon_i(T,\lambda)$. The system $\{F_1, F_2\}$ has no external radiation.

The surface density of the resulting radiation from the surface F_1 (the resulting heat flow from F_1) is well-characterized by a known function $q(M_1,\lambda)$. We consider all the functions involved in the process of the radiation exchange between the surfaces F_1 and F_2 to be continuous and write the integral equation establishing a precise relationship between the given quantities $T(M_1), q(M_1,\lambda)$ and the unknown function $T(M_2)$ (see Rusin and Leonov (1987, 1991)):

$$\begin{aligned}&\iint_{F_2} K(M_1,N_2)\varepsilon_\Sigma[T(N_2)][T(N_2)/T_0]^4 dS_{N_2}\\ &\qquad + \iint_{F_2}\iint_{F_1} K(M_1,N_2)K(N_2,N_1)B\big[T(N_2),N_1\big]dS_{N_2}dS_{N_1}\\ &= u(M_1) \end{aligned}\tag{1}$$

In this context, we take for granted the natural condition $T(M_2) \geq T_{min} > 0$, where T_{min} is a known constant. In studying equation (1) we keep the following notions: N_1 and N_2 are points on the surfaces F_1 and F_2, dS_{N_i} is a surface element of F_i, $K(M_i,N_j)dS_{N_j}$ for $i,j = 1,2$ is the so-called elementary angle coefficient of the cell dS_{M_i} with regard to the cell dS_{N_j} (see Rusin and Peletskiĭ (1987)). The functions $K(M_i,N_j)$ are determined by the geometry of the surfaces F_1 and F_2. The function $\varepsilon_\Sigma(T)$, which indicates the integral hemispherical emittivity of the surface F_2, is also given. It can be determined by experiment and is well-known for many materials. The constant T_0 corresponds to a known average temperature at the surface F_1. The operator $B[T(N_2),N_1]$ assigns to every function $T(N_2)$ the values

$$\begin{aligned} B[T(N_2),N_1] &= \sigma_0^{-1}T_0^{-4}\int_0^\infty r_2\big[T(N_2),\lambda\big]\\ &\quad\times\Big\{E_0\big[T(N_1),\lambda\big] - r_1\big[T(N_1),\lambda\big]\\ &\quad\times a_1^{-1}\big[T(N_1),\lambda\big]q(N_1,\lambda)\Big\}d\lambda \end{aligned}$$

where the function $E_0(T,\lambda)$ is taken from the well-known Planck formula and σ_0 is the Stefan–Boltzmann constant (see Rusin and Peletskiĭ (1987)). The function $u(M_1)$ in the right-hand side of equation (1) can always be calculated from the known quantities $K, T(M_1), T_0, q(M_1,\lambda), r_1(T,\lambda)$ and

$a_1(T, \lambda)$:

$$u(M_1) = \left[\frac{T(M_1)}{T_0}\right]^4 - \iint\limits_{F_1} K(M_1, N_1)\left[\frac{T(N_1)}{T_0}\right]^4 dS_{N_1}$$

$$-\sigma_0^{-1}T_0^{-4}\int\limits_0^\infty q(M_1,\lambda)a_1^{-1}[T(M_1),\lambda]d\lambda + \sigma_0^{-1}T_0^{-4}\iint\limits_{F_1} K(M_1,N_1)$$

$$\times\left\{\int\limits_0^\infty r_1[T(N_1),\lambda]a_1^{-1}[T(N_1),\lambda]q(N_1,\lambda)d\lambda\right\}dS_{N_1}$$

The nonlinear integral equation (1) related to the temperature field $T(N_2)$ at the heater can be represented in the operator form

$$A[M_1, T(N_2)] = u(M_1) \quad u(M_1) \in L_2(F_1) \quad T(N_2) \in D \tag{2}$$

where A is a nonlinear integral operator from the set of solutions $D = \{T(N_2) \in C(F_2): T(N_2) \geq T_{min}\}$ into $L_2(F_1)$. It is specified by the left-hand side of equality (1). The operator equation (2) falls within the category of ill-posed problems, since its solutions do not necessarily exist on the set D in the case when a given field $T(M_1)$ is physically unrealizable. This equation may be also unstable with respect to perturbations of the operator A due to the experimental quantities $\varepsilon_\Sigma, r_1, a_1$ as well as with respect to inevitable errors produced by discretization of operator (2) that will be needed in the process of its numerical solution. Therefore, we are able to set up the problem of searching for Ω-optimal quasisolutions to equation (2) on the set D in the framework of Section 3.1 in Chapter 3.

Of special interest is the functional

$$\Omega[T(N_2)] = \iint\limits_{F_2} \varepsilon_\Sigma[T(N_2)]\left[\frac{T(N_2)}{T_0}\right]^4 dS_{N_2} \tag{3}$$

From a physical point of view it corresponds to the value of the total energy radiated from the heater's surface (see Rusin and Leonov (1991)). The temperature field $\bar{T}(N_2)$, being an Ω-optimal quasisolution to equation (2) on the set D for that functional, not only satisfies equations (1) and (2) in the best way in the sense of minimal $L_2(F_1)$-norm, but also provides minimal power expenditures during the work of such a heater.

To find Ω-optimal quasisolutions to equation (2) on the set D one can employ the algorithms of Chapter 3 including the algorithm of the generalized discrepancy principle. We have applied the program realization described in Sections 2.14 in Chapter 2 and 4.5 in Chapter 4 in which the condition $T(N_2) \geq T_{min}$ was taken into account. In this way we have solved many problems concerning the design of optimal distribution of a temperature

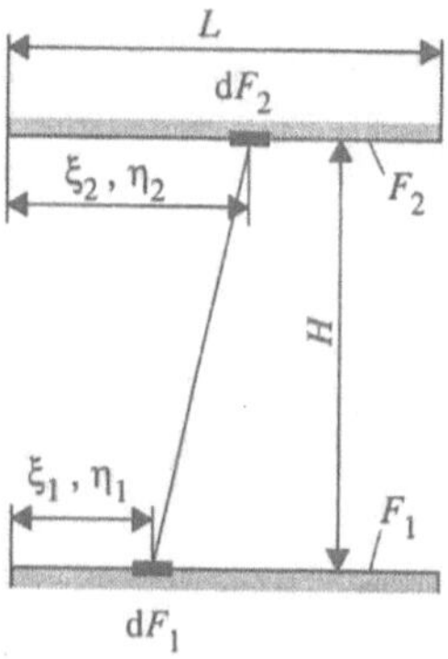

Figure 7.5 The scheme of the radiating system.

field on a heater for maintaining the available values $T(M_1)$ and $q(M_1, \lambda)$ on an object. We give below an elementary example.

Let the surfaces F_1 and F_2 represent two parallel infinite bands of width L perpendicular to the plane of the figure (Fig. 7.5) with spacing H. The material of the bands is polished tungsten. The dependence $\varepsilon(T, \lambda)$ for this material in the range of wavelengths λ from 0.3 to 5 mcm is well-known. The function $\varepsilon_\Sigma(T)$ for tungsten can be determined by experiment. This example devotes to the problem of finding the temperature field at the heater enabling to effect the constant temperature field $T(M_1) = T_1 = 1000$ K on the object for $q(M, \lambda) = 0$. We agree to consider $T(M_2) \geq T_{min} = 300$ K. In this case equation (1) takes the form

$$\int_0^1 K(\xi_1, \eta_2)\varepsilon_\Sigma\big[T(\eta_2)\big]\big[T(\eta_2)/T_1\big]^4 d\eta_2 + \int_0^1 K(\xi_1, \eta_2)\varphi(\eta_2)B\big[T(\eta_2)\big]d\eta_2 = 1 \equiv u(\xi_1)$$

$$T(\eta_2)/T_1 \geq 0.3 \tag{4}$$

where

$$B\big[T(\eta_2)\big] = 1 - \frac{1}{\sigma_0 T_1^4}\int_0^\infty \varepsilon\big[T(\eta_2), \lambda\big]E_0(T_1, \lambda)d\lambda$$

$$K(\xi, \eta) = \gamma^2\big[(\xi - \eta)^2 + \gamma^2\big]^{-3/2}/2 \qquad \gamma \equiv H/L$$

$$\varphi(\eta_2) = \int_0^1 K(\eta_2, \eta_1)d\eta_1$$

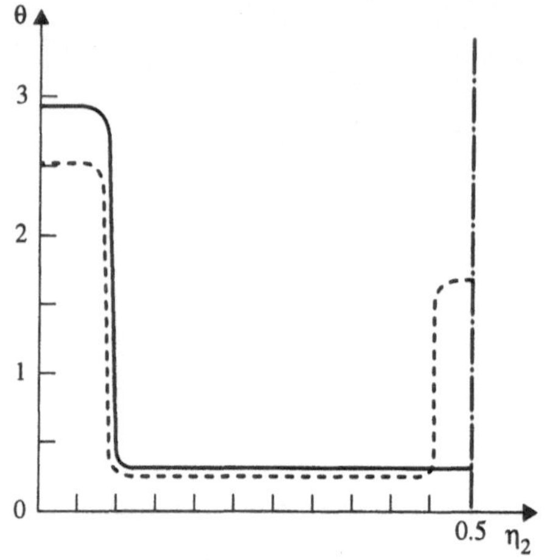

Figure 7.6 The distribution of the relative temperature $\theta(\eta_2)$ on the heater's surface F_2 for different γ: $\gamma = 1$ — ; $\gamma = 0.5$ - - -.

The quantities ξ_1, η_1 and η_2 are dimensionless coordinates on the surfaces F_1 and F_2 (normed to L). A reasonable form of functional (3) is

$$\Omega[T(\eta_2)] = \int_0^1 \varepsilon_\Sigma[T(\eta_2)] \left[\frac{T(\eta_2)}{T_1}\right]^4 d\eta_2$$

In Fig. 7.6 the results of determining Ω-optimal quasisolutions to equation (4) $\theta(\eta_2) \equiv \bar{T}(\eta_2)/T_1$ are shown for various values of γ which indicate the relative spacing between the surfaces. It is worth noting that the requirement $T(M_1) = T_1$ is wittingly unrealizable for this system. In this context, it is of considerable interest to investigate the behaviour of the equation (4) discrepancy on Ω-optimal quasisolutions $\bar{T}(\eta_2)$ thus obtained. So, we are interested in the diagram of the left-hand side of equation (4) with $T(\eta_2) = \bar{T}(\eta_2)$ as the following function of ξ_1:

$$\begin{aligned}\psi(\xi_1) &= \int_0^1 K(\xi_1, \eta_2)\varepsilon_\Sigma[\bar{T}(\eta_2)] \left[\frac{\bar{T}(\eta_2)}{T_1}\right]^4 d\eta_2 \\ &\quad + \int_0^1 K(\xi_1, \eta_2)\varphi(\eta_2) B[\bar{T}(\eta_2)] d\eta_2\end{aligned}$$

When this diagram is compared with that of the required right-hand side of the equation $u(\xi_1) = 1$ (see Fig. 7.7), it is apparent that the difference is about 2–4% depending on particular values of γ.

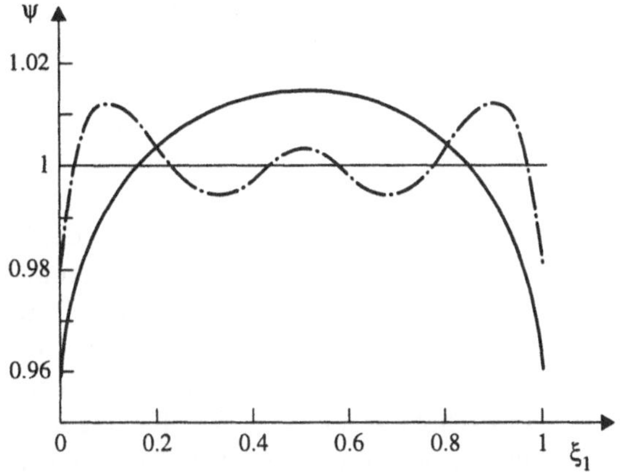

Figure 7.7 Comparison of the dependences $\psi(\xi_1)$ corresponding to the functions $\theta(\eta_2)$ for different γ: $\gamma = 1$ —; $\gamma = 0.5$ – · –; $u(\xi_1) \equiv 1$.

The results of the preceding calculations, as numerous others, demonstrate that various heaters with zonal structure of the temperature field are needed to provide the constant temperature $T(M_1)$ on the object. In such cases the width of zones, their location at the heater and their temperature can be determined in a similar manner as we did before.

7.4 Restoration of the absorption lines local profiles in spectra of Ap-stars

Following the original works of Goncharskiĭ, Stepanov *et al.* (1977, 1982) and Goncharskiĭ, Cherepashchuk and Yagola (1985) we now consider an interesting nonlinear inverse problem of astrophysics. Many stars, including the so-called Ap-stars, manifest periodic variations in the absorption lines profiles of some elements. Such variations can be explained by the fact that chemical elements possess a nonuniform distribution on the surface of a rotating star and, hence, by the difference in the absorption lines profiles in various regions on the star's surface. The fact that the changes of profiles are produced by nonhomogeneous chemical compositions rather than by different physical conditions on the star's surface, can be easily checked by constancy or very small variations of the spectral structure of continuous radiation of a star and its brightness, which are mainly defined by physical conditions in the star's atmosphere. In addition, the synchronism frequently observed in variations of profiles of distinct elements, even if they have close excitation and ionization potentials, contribute significantly to this verification.

Quite often, similar stars possess essential magnetic fields varying with the same periods as the line profiles. Apparently, the nonhomogeneity is produced by the strong magnetic field of a star. This property adds interest and help in understanding of phenomena and localizing the chemical composition anomalies for subsequent comparisons with the distribution of the magnetic field.

The main goal of our studies is the problem of reconstructing the absorption lines local profiles in which it is required to determine from the available profiles of the absorption lines given at different times those of lines observed at a fixed point on the star's surface. Data processing of these profiles by classical methods allows us to calculate the number of absorbing atoms or ions at each point on the surface. A common setting of this problem is given in Khokhlova (1976). Special attention is paid to the absorption lines profiles in the star's spectrum observed from the Earth. We use the notion of the radiation intensity $I(M;\theta,\lambda)$ at a point M on the star's surface at a frequency $\nu = c/\lambda$ (c is the light velocity) at an angle θ to the normal expressed in terms of the energy dE being transferred by radiation across a unit surface area at the angle θ to its normal in the solid angle $2\pi\sin\theta d\theta$ in the wavelength band $[\lambda, \lambda + d\lambda]$ during the period $[t, t + dt]$. These variables are related by

$$dE = 2\pi I(M;\theta,\lambda)\sin\theta d\theta\cos\theta d\lambda dt \tag{1}$$

By means of the radiation intensity defined in such a way, we deduce that the energy coming to the Earth from the entire visible surface of a star is proportional to the quantity

$$F_\lambda^+ = \iint I(M;\theta,\lambda)\cos\theta dM \tag{2}$$

where dM is a surface element on the sphere; θ is the angle between the normal to the star's surface at the point M and the viewing line and integration is accomplished over the entire visible surface of the star. If we introduce the spherical system of coordinates (θ, φ) whose axis is directed along the viewing line, then $dM = \sin\theta d\theta d\varphi$ and the visible surface is chosen so that the condition $\cos\theta > 0$ holds.

For stars rotating rapidly enough, in particular for Ap-stars, it should be taken into account that Doppler's shift of the radiated light may differ at different points on the star's surface and may attain the values comparable with the line width. Thereby the argument λ of the function $I(M;\theta,\lambda)$ in the integrand should be replaced by $\lambda + \Delta\lambda_D(M)$ with $\Delta\lambda_D(M)$ being Doppler's shift of radiation at a point M on the rotating star's surface.

We will distinguish the radiation intensity in the continuum, that is, far from the center of line (or the intensity radiation when there are no absorbing atoms) $I_0(M;\theta,\lambda)$ and the radiation intensity in the line which depends naturally on the chemical compositions of absorbing atoms in the atmosphere.

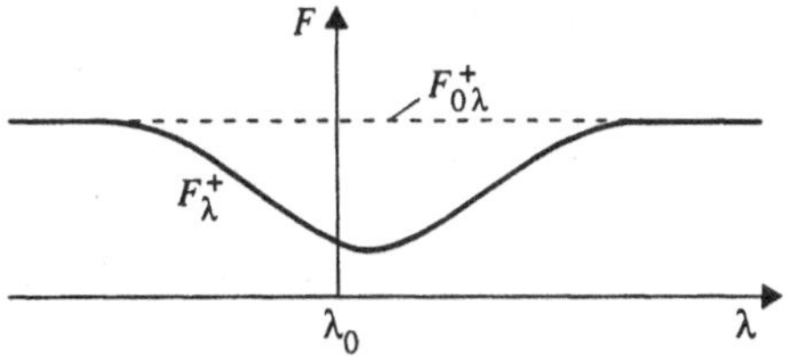

Figure 7.8 The graph of the function $F(\lambda)$.

The spectrum monitored on the Earth is schematically shown in Fig. 7.8. Here $F^+_{0\lambda}$ is the flow due to the continuum and F^+_λ is the flow produced by the absorption in the line. Since the absorption in the Earth's atmosphere is one and the same for the line spectrum and the continuum, the quantity

$$R(\lambda) = 1 - \frac{F^+_\lambda}{F^+_{0\lambda}} = 1 - \frac{\iint\limits_{\cos\theta>0} I(M;\theta,\lambda+\Delta\lambda_D(M))\cos\theta dM}{\iint\limits_{\cos\theta>0} I_0(M;\theta,\lambda+\Delta\lambda_D(M))\cos\theta dM} \tag{3}$$

which does not depend on the state of the Earth's atmosphere, is commonly used to remove the dependence of this absorption. We call it the profile of the line.

The functions $I_0(M;\theta,\lambda)$ and $I(M;\theta,\lambda)$ are solutions to equations of the radiation transfer including the absorption and re-radiation coefficients in the line as well as the absorption and re-radiation coefficients in the continuum as certain functions of the depth in the atmosphere. Coefficients in the continuum are defined mainly by the dependence of hydrogen and its ions on the depth or by the atmosphere model, as it were.

Because the measured wavelength range is rather narrow, the continuum intensity is practically constant and may be taken to be independent of λ. Within the approximation so constructed, we may assume, in addition, that the continuum intensity is irrelevant to a particular point M on the star's surface, since physical conditions on the star may be viewed as homogeneous. In accordance with what has been said, we might agree with $I_0(M;\theta,\lambda) = I_0(\theta)$. On the other hand, the dependence of the continuum intensity on the angle is known as the limb darkening and takes the form $I_0(\theta) = I_0 \cdot u_1(\theta)$, where $u_1(\theta)$ is a linear function of $\cos\theta$ (see, for example, Aller (1961)). This law of limb darkening emerging from theoretical reasons is consistent with available variations of the radiation intensity occurring from the center of the solar disk to its limb as well as with interpretations of brightness curves of double darkened systems. The dependence of the coefficient u_1 upon the temperature in the representation of the function $u_1(\theta) = 1 - u_1 + u_1\cos\theta$ is carefully analyzed, while the star's temperature can be recovered from the available information on its spectral class.

We know from the paper by Khokhlova (1976) that the function

$$r(M,\lambda) = 1 - \frac{I(M;0,\lambda)}{I_0(M;\lambda)} \tag{4}$$

can always be expressed more precisely by appeal to Minnaerth's formula:

$$r(M,\lambda) = \left\{ \frac{1}{x(M,\lambda)} + \frac{1}{R_c(M)} \right\}^{-1} \tag{5}$$

where $R_c(M)$ is the central depth of the line, $x(M,\lambda) = \tau(M)k(\lambda)$; $k(\lambda)$ being a known function of λ characterizes the so-called Foygt profile. The coefficient $\tau(M)$ depending on a point on the star's surface is defined by the relative content of the absorbing element in the atmosphere.

The dependence of the radiation intensity in a line $I(M;\theta,\lambda)$ upon the angle θ becomes rather complicated. The absorption lines may intensify to the star's limb and even turn to emission ones. The reader can encounter the phenomenon where the central parts of a line relax to the limb, while the wings of the same line may intensify. However, for many lines these variations from the center to the limb are negligible. Bearing in mind the results obtained in Khokhlova (1976) we will assume in the sequel that the appropriate dependence can be written as

$$1 - \frac{I(M;\theta,\lambda)}{I_0(M;\theta)} = r(M,\lambda)u_2(\theta) \tag{6}$$

where $u_2(\theta)$ is a linear function with respect to $\cos\theta$: $u_2(\theta) = 1 - u_2 + u_2\cos\theta$. A numerical parameter u_2 can be found in many ways, for instance, by numerical integration of the transfer equation of an atmosphere model close to the atmosphere of the star in view. For profiles with small variations from the center to the limb, the coefficient is taken to be zero.

When physical conditions happen to be homogeneous, the parameters τ and R_c involved in Minnaerth's formula (5) become dependent. Indeed, the variables R_c and τ correspond to the central depth and width of the line. So, for small R_c and large τ formula (5) describes a shallow and broad line and, conversely, for large R_c and small τ it corresponds to a deep and narrow line. If the temperature and other physical parameters of atmosphere are held fixed, such profiles cannot exist simultaneously. Along with increasing the absorbing element content, the line becomes deeper until its center is saturated and, after this, it only widens. In this case the central depth under saturation is a function of the surface temperature. As shown in Pavlova and Khokhlova (1980), in the study of the line profiles obtained as a result of solving the transfer equation, the parameters R_c and τ are dependent and the dependence of the central depth of the value τ can be approximated as follows:

$$R_c = h(\tau) = c_1\big[1 - \exp\{-c_2\tau\}\big] \tag{7}$$

The parameters c_1 and c_2 are suitably chosen for each star and each line on the basis of an atmosphere model capable of describing a star.

Finally, the absorption lines profiles in the spectrum of the integral radiation of the whole star's disk observed from the Earth can be determined via the function $\tau(M)$ by the formula

$$R(\lambda) = \frac{1}{N} \iint\limits_{\cos\theta>0} f\big[\tau(M), \lambda + \Delta\lambda_D(M)\big] u_1(\theta) u_2(\theta) \cos\theta dM \tag{8}$$

with

$$N = 2\pi \int\limits_0^{\pi/2} \cos\theta u_1(\theta) \sin\theta d\theta \tag{9}$$

where θ is the angle between the normal at a point M and the viewing line, $\Delta\lambda_D(M)$ is Doppler's shift of radiation due to the rotation of the star and

$$f(\tau, \lambda) = \left\{ \frac{1}{h(\tau)} + \frac{1}{\tau k(\lambda)} \right\}^{-1} \tag{10}$$

It is worth noting here that in the process of the star's rotation the profiles $R(\lambda)$ may vary. The traditional way of covering this is to have occasion to use $R(\lambda, \omega t)$, where t is the time and ω is the angular rotation velocity. With these assumptions, equation (8) reduces to

$$R(\lambda, \omega t) = \iint f\big[\tau(M), \lambda + \Delta\lambda_D(M, \omega t)\big] H(M, \omega t) dM \tag{11}$$

If we regard the latitude Φ and the longitude l as the coordinates on the star's surface introduced above, then a surface element is equal to $dM = \cos\Phi d\Phi dl$ and the function $f(\tau, \lambda)$ is the same as we considered before. In the new coordinates,

$$\Delta\lambda_D(\Phi, l, \omega t) = v_{eq} \sin i \cos\Phi \cos(l - \omega t) \lambda_0 / c \tag{12}$$

$$H(\Phi, l, \omega t) = \left\{ \frac{1}{N} u_1(\theta) u_2(\theta) \cos\theta \cos\Phi\colon\ \cos\theta > 0; 0\colon\ \cos\theta \le 0 \right\} \tag{13}$$

$$\cos\theta = \cos i \sin\Phi + \cos\Phi \sin i \cos(l - \omega t) \tag{14}$$

Here v_{eq} is the equatorial velocity of the star's rotation and i is the angle of the slope of the star's rotation axis to the viewing line.

The Foygt profile $k(\lambda)$ characterizing the dependence of the profile upon the wavelength is a convolution of two functions describing the profile of the line widened by the pressure (collisions) and Doppler's shift in the process of microturbulent motions in the star's atmosphere:

$$k(\lambda) = \frac{\varkappa_l}{\varkappa_d} \frac{1}{\pi} \int\limits_{-\infty}^{\infty} \frac{e^{-t^2} dt}{\varkappa_l^2 / \varkappa_d^2 + (t - \lambda/\varkappa_d)^2} \tag{15}$$

Here $\varkappa_l$ is the quantity of the Lorentz widening by the pressure. The parameter $\varkappa_d$ is of formal character to a considerable extent and indicates the value of the line widening by Doppler's shift. After scrutinizing an approved atmosphere model, one is to choose this parameter so that Minnaerth's formula will adequately describe the absorption lines profiles for various concentrations of absorbing atoms.

This is acceptable if we are willing to recover the function $\tau(M)$ from the available function $R(\lambda, \omega t)$ which represents profiles of the line measured at different phases. In each such case we may treat (11) as a nonlinear integral equation related to the function $\tau(M)$. The restriction $\tau(M) \geq 0$ on the function $\tau(M)$ arises naturally.

In fact, a knowledge of the function $\tau(M)$ means a knowledge of the absorption lines local profiles $r(M, \lambda)$ and hence a knowledge of the chemical composition of the element responsible for the appropriate absorption line at a point M on the star's surface.

It remains to make a remark necessary for some parameters emerging from equation (11) and having been previously found from independent observations. First of all it is concerned with parameters v_{eq} and $\sin i$. The quantity $v_{eq} \sin i$ can be evaluated by measuring the half-width $\Delta\lambda_{0.5}$ of some constant lines for which the widening is mainly due to the star's rotation. Then $v_{eq} \sin i \approx 1.25c\Delta\lambda_{0.5}/(2\lambda)$, where c is the light velocity. Moreover, we may assume, as usual, that any star from this class is of radius R, which is about 2–4 R_S, where R_S is the Sun's radius, and that the estimation is based on the available information about the star's spectral class. Therefore, $v_{eq} = 2\pi R/P$, where P is the period of the star's rotation. The problem under consideration was formulated in such a form for the first time by Goncharskiĭ, Stepanov *et al.* (1977).

It is convenient to introduce several definitions which will be needed in subsequent reasonings. By an equivalent line width, we mean the quantity (possibly dependent on the time)

$$W(\omega t) = \int_0^\infty R(\lambda, \omega t) d\lambda = \iint \xi(M) H(M, \omega t) dM \qquad (16)$$

where

$$\xi(M) = \int_0^\infty r(M, \lambda) d\lambda \qquad (17)$$

is a local equivalent line width. We define also an effective Doppler's shift by the integral

$$\begin{aligned} Z(\omega t) &= \int_0^\infty R(\lambda, \omega t)(\lambda - \lambda_0) d\lambda \\ &= \iint \xi(M) \Delta\lambda_D(M, \omega t) H(M, \omega t) dM \qquad (18) \end{aligned}$$

where λ_0 is the center of the line in question.

First we consider methods for studying the chemical composition of normal stars with uniform distribution of the absorbing substance on the surface, what means that $r(M, \lambda)$ and $\xi(M)$ are both independent of M. In this case only observed equivalent widths W are usually needed in investigations. When the energy distribution in the radiation continuum of such stars, its hydrogen and the helium absorption lines are explored, one can find the following atmosphere parameters: the temperature T_{ef}, the acceleration of gravity g and the ratio of helium and hydrogen. These parameters permit one can carry out the descriptions of an atmosphere model by means of which it is possible to recover the dependence of the equivalent width of the absorption line on the relative concentration of the absorbing element in the atmosphere known as the increase curve while solving the transfer equation. The use of the increase curve allows us to determine the amount of the absorbing substance if equivalent widths of its absorption lines are known. The method for recovering the chemical composition on the basis of an atmosphere model is termed the atmosphere model method. This method proves to be useful in solving the most important problems of astrophysics. This is especially true for the problem of abundance of chemical elements studied in detail in Aller (1961).

For stars with nonhomogeneous chemical compositions at the surface it is desirable to split up the problem of determining the distribution of the chemical element responsible for a given absorption line into the following ones:

(a) the problem of reconstructing local equivalent widths or the lines profiles;

(b) the problem of finding the local chemical composition from equivalent widths thus obtained or the lines profiles.

Problem (b) can be solved by using the atmosphere model method described above. Let us turn to the questions of possibilities of applying the regularization method of ill-posed problems to the inverse problem of reconstructing the absorption lines local profiles. Retaining the above notations, we proceed to consider problem (a).

In studying equation (11) a reasonable form of the smoothing functional is

$$\begin{aligned} M^{\alpha}[\tau] &= \iint \alpha\left\{\left(\frac{d\tau}{d\Phi}\right)^2 + \left(\frac{d\tau}{dl}\right)^2\right\} d\Phi dl \iint \Big\{R(\lambda, \omega t) \\ &\quad - \iint f\big[\tau(M), \lambda + \Delta\lambda_D(M, \omega t)\big] \\ &\quad \times H(M, \omega t) dM\Big\}^2 d\lambda d\omega t \end{aligned} \qquad (19)$$

Assuming that the function $\tau(M)$ belongs to the space W_2^1 comprising the functions defined on the star's surface, it is required to minimize functional (19) on the set of all nonnegative functions $\tau(M)$. We choose the regularization parameter α by means of the alternative principle of Section 3.5 in Chapter 3. Since the star's regions with $\Phi \in [-\pi/2, -i]$ are not seen from the Earth, the function $\tau(M)$ is sought on the set $\Phi \in [-i, \pi/2], l \in [-\pi, \pi]$. The function $R(\lambda, \omega t)$ representing the available experimental data is supposed to be known for $\lambda \in (\lambda_0 - \Delta\lambda/2, \lambda_0 + \Delta\lambda/2), \lambda \in [-\pi, \pi]$. Thus, in expression (19) for $M^\alpha[\tau]$ integration is accomplished over the set $\Phi \in [-i, \pi/2], l \in [-\pi, \pi]$.

One might reasonably try to construct a finite difference approximation of the problem by introducing on the star's surface in the coordinates $\Phi \in [-i, \pi/2], l \in [-\pi, \pi]$ a rectangular grid $\{\Phi_p, l_q\}$ consisting of $N \times M$ nodes. Each such node is supposed to be the center of a cell on the star's surface, the value of the function $\tau(\Phi, l)$ being constant in every cell: $\tau(\Phi, l) = \tau_{pq}, p = 1, \dots, N; q = 1, \dots, M$. Furthermore, we choose an equidistant grid with respect to λ consisting of N_λ nodes including the line center and denote them by $\lambda_k, k = 1, \dots, N_\lambda$. We assume also that profiles are observed at $N_{\omega t}$ phases $(\omega t)_l, l = 1, \dots, N_{\omega t}$. With the notation $R_{kl} = R[\lambda_k, (\omega t)_l]$, the integrals with respect to λ and ωt from expression (19) are approximated by the rectangles formula on those grids. Instead of l, we are working with a new variable $\psi = l - \omega t$, which ranges in (19) from $-\pi$ to π like the usual variable l. But according to (13)–(14) we might have $H(\Phi, \psi) = 0$ if $\cos\theta = \sin\Phi \cos i + \cos\Phi \sin i \cos\psi \le 0$, which means that $H(\Phi, \psi) \ne 0$ only if $-\Gamma(\Phi) \le \psi \le \Gamma(\Phi)$, where $\Gamma(\Phi) = \arccos(\mathrm{tg}\Phi \mathrm{ctg} i)$. We refer to equidistant grids $\{\Phi_i\}$ and $\{\psi_j\}$ in Φ and ψ with spacings h_1 and h_2, respectively. The integral part of $[\Gamma(\Phi)/h_2]$ will be denoted by $L(\Phi)$. Let $p(i, j, l)$ and $q(i, j, l)$ be the coordinates of the cell center on the star's surface from the rectangle $\Phi \in [-i, \pi/2], l \in [-\pi, \pi]$ enclosing the point $(\Phi_i, \psi_j + (\omega t)_l)$. According to formula (15) Foygt's profile is the convolution of two functions. It would be erroneous to find it every time because the operation of convolution is accompanied by cumbersome calculations. In mastering difficulties involved, we propose the following algorithm designed for computing the function $k(\lambda)$. Observe that in the ranges of the parameters $\varkappa_l$ and $\varkappa_d$ the function $k(\lambda)$ can be approximated more precisely by Matveev's formula (1972):

$$\begin{aligned} k(\lambda) &= \frac{\sqrt{\log 2}}{\sqrt{\pi}\varkappa_v}(1-\xi)\exp\{-\eta^2 \log 2\} \\ &\quad + \frac{1}{\pi\varkappa_v}\frac{1}{1+\eta^2} - \frac{\xi(1-\xi)}{\pi\varkappa_v}\left(\frac{1.5}{\log 2} + 1 + \xi\right) \\ &\quad \times \left[0.066\exp\{-0.4\eta^2\} - \frac{1}{40 - 5.5\eta^2 + \eta^4}\right] \end{aligned} \tag{20}$$

$$\begin{aligned} \varkappa_v &= 0.5\left(\varkappa_l + \sqrt{\varkappa_l^2 + 4\varkappa_d^2}\right) + 0.05\varkappa_l \\ &\quad \times \left(1 - \frac{2\varkappa_l}{\varkappa_l + \sqrt{\varkappa_l^2 + \varkappa_d^2}}\right) \\ \eta &\equiv (\lambda - \lambda_0)/\varkappa_v \qquad \xi \equiv \varkappa_l/\varkappa_v \end{aligned}$$

With the aid of (20) we can fill in the table of the values of $k(\lambda)$ on an equidistant grid (with respect to λ) consisting approximately of 300–600 points so that it will cover all the values of λ arising when calculating (20). The value of $k(\lambda)$ at an arbitrary point can be obtained by the linear interpolation. Finally, one of the finite difference approximations to functional (19) is given by

$$\widehat{M}^{\alpha}[\tau] = \alpha\widehat{\Omega}[\tau] + \widehat{\Phi}[\tau] \tag{21}$$

where

$$\begin{aligned} \widehat{\Phi}[\tau] &= \sum_{k=1}^{N_\lambda}\sum_{l=1}^{N_{\omega t}} w_{kl}\Bigg\{R_{kl} \\ &\quad -\sum_{i=1}^{N}\sum_{j=-L(\Phi_i)}^{L(\Phi_i)} h_1 h_2 H(\Phi_i,\psi_j) \\ &\quad \times f\left[\tau_{p(i,j,l)q(i,j,l)}, \lambda_k + \Delta\lambda_D(\Phi_i,\psi_j)\right]\Bigg\}^2 \end{aligned} \tag{22}$$

$$\begin{aligned} \widehat{\Omega}[\tau] &= \sum_{j=1}^{M}\sum_{j=2}^{N}(\tau_{ij} - \tau_{i-1,j})^2 \\ &\quad + \sum_{i=1}^{N}\Bigg\{\sum_{j=2}^{M}(\tau_{ij} - \tau_{i,j-1})^2 \\ &\quad + (\tau_{i1} - \tau_{iM})^2\Bigg\} \end{aligned} \tag{23}$$

Here w_{kl} are the weights introduced for the following reasons: the accuracy in measuring $R(\lambda, \omega t)$ and the number of measurements of $R(\lambda, \omega t)$ may differ at different phases. Usually the profiles corresponding to each phase are known with a high degree of accuracy in the center of the line and with a lower degree of accuracy on the wings due to the blending effect.

So, the problem of reconstructing the absorption lines local profiles leads to minimizing the functional $\widehat{M}^{\alpha}[\tau]$ on a finite-dimensional space of the dimension $N \times M$ under the restrictions $\tau_{pq} \geq 0$ by a proper choice of the regularization parameter in complete agreement with the alternative principle.

For this reason we have been trying to adapt the projection method of conjugate gradients for minimizing $\widehat{M}^\alpha[\tau]$ (see, for instance, Karmanov (1975), Pshenichny and Danilin (1975), Vasilyev (1980)). The iteration process can be arranged in the following way. The initial approximation is viewed as a vector $\tau^{(0)}$ with nonnegative components which will serve as a necessary background for constructing the gradient $g^{(i)} = \mathrm{grad}\widehat{M}^\alpha[\tau^{(i)}]$ at a given point $\tau^{(i)}$ of the minimizing sequence. It is not difficult to derive analytic expressions for the gradient of $\widehat{M}^\alpha[\tau]$. With a knowledge of the gradient's direction at the previous step $g^{(i)}$ and the direction of descent at the previous step $h^{(i)}$, it is possible to find a new direction of descent by appeal to the formula

$$h^{(i)} = g^{(i)} + \gamma_i h^{(i-1)} \tag{24}$$

where

$$\gamma_i = \frac{\left(g^{(i)} - g^{(i-1)}, g^{(i)}\right)}{\left(g^{(i-1)}, g^{(i-1)}\right)} \tag{25}$$

This is the method of conjugation known as the Polak–Ribiére version (see Polak (1971)). The next step is to construct the one-parameter set of the elements

$$\tau_\lambda = P\left(\tau^{(i)} - \lambda h^{(i)}\right) \qquad \lambda \geq 0 \tag{26}$$

where P is the projector of an $N \times M$-vector onto the first octant and the problem of the one-dimensional minimization of the functional $\widehat{M}^\alpha[\tau]$ on that set is to be solved. The point at which the function $\widehat{M}^\alpha[\tau]$ attains its minimum is adopted as the next element $\tau^{(i+1)}$ of the minimizing sequence. For solving the relevant one-dimensional minimization problem, it is possible to use the quadratic approximation of the function

$$\theta(\lambda) \equiv \widehat{M}^\alpha\left[P(\tau^{(i)} - \lambda h^{(i)}\right] - \widehat{M}^\alpha\left[\tau^{(i)}\right] \tag{27}$$

at three points $\lambda = 0, \lambda_n, 2\lambda_n$ and then to adopt

$$\lambda_{min} = \lambda_n \frac{1}{2}\frac{\theta_2 - 4\theta_1}{\theta_2 - 2\theta_1} \qquad \theta_1 \equiv \theta(\lambda_n) \quad \theta_2 \equiv \theta(2\lambda_n) \tag{28}$$

as an approximate solution (one should refine this point additionally if necessary). After every $N \times M$ steps the method must be 'revised' by equating γ to zero in formula (24). Moreover, the algorithm will be changed if the inaccuracy in solving the one-dimensional minimization problem may yield $\left(h^{(i)}, g^{(i)}\right) \leq 0$ at the current step.

The algorithm presented above was realized by Dr V.V. Stepanov as a program package in Fortran. He has carried out computations for modelling problems given below on the basis of experimental information submitted by V.L. Khokhlova. The influence of various parameters involved in the problem statement for the profiles of the observed spectral lines has been investigated. It is concerned with coefficients of limb darkening in the continuum and the line u_1, u_2, the angle between the axis of the star's rotation and the viewing line i, the equatorial rotation velocity v_{eq}, the parameters of Foygt's profile $\varkappa_l, \varkappa_d$ as well as with parameters c_1, c_2 of the dependence $h(\tau)$. It has been shown that the influence of parameters $u_1, u_2, \varkappa_l, \varkappa_d, c_1$ and c_2 on the integral profiles of lines is not significant and, in order to solve the inverse problem, it is enough to obtain rough estimates of these parameters. In most cases the careful analysis of available experimental data may be of help in achieving this aim.

The predicted variations of the parameters i and v_{eq} result in variations of the observed integral profiles to a considerable extent. As shown, the line profiles of $R(\lambda, \omega t)$ may be changed not only quantitatively, but also qualitatively for the varying parameter v_{eq}. This obstacle necessitates evaluating the parameters i and v_{eq} more accurately before proceeding.

As has been mentioned before, the domain of $\tau(M)$ is different for distinct values of i. The greater is the angle i, the larger is the region visible from the Earth. When $i = \pi/2$ the entire star's surface can be observed, but in this case $\tau(M)$ cannot be recovered uniquely. For i about zero Doppler's shift of the radiation from points on the star's surface is small and, therefore, it is impossible to recover the distribution of $\tau(M)$. From such reasoning it seems clear that it is extremely difficult to reconstruct $\tau(M)$ accurately on the boundary of the region and in the vicinity of poles.

Below we quote the results of numerical solution of modelling problems. In all of these cases we used the uniform distribution $\tau(M) = 0.7$ as an initial approximation for constructing a minimizing sequence for a suitably chosen smoothing functional. A very small content of absorbing elements corresponds to this value and practically it does not generate any line. In further processing of local profiles, the distribution of local equivalent widths will be of great interest and, therefore, it is convenient to compare the results in terms of a local equivalent width.

The map of the distribution of local equivalent widths is plotted in Fig. 7.9. This distribution has been obtained by applying the method described above to the line profiles in the case where there are three rectangular regions on the star's surface with a higher content of the absorbing element. The contours of those regions representing an exact solution are shown in Fig. 7.9. It is easily seen that the location of the spots is determined in longitude rather accurately. The sizes of the spots in latitude are enlarged and the line intensity is somewhat relaxed with respect to the initial one. It is worth noting here that the distribution obtained permits us to describe the observed profiles of the lines with accuracy up to 0.01, which corresponds to errors in specifying the experimental information. These profiles are depicted in Fig. 7.10.

Variations of a local equivalent width on the surface correspond to variations of the relative content of an absorbing element by some orders. For

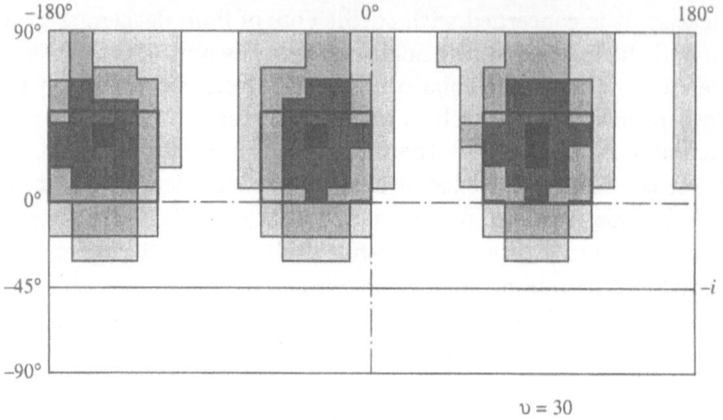

Figure 7.9 The map of distribution of local equivalent widths (model 1): (1) 0.03–0.08 Å; (2) 0.08–0.13 Å; (3) 0.13–0.18 Å; (4) 0.18–0.23 Å; (5) contours of the initial distribution.

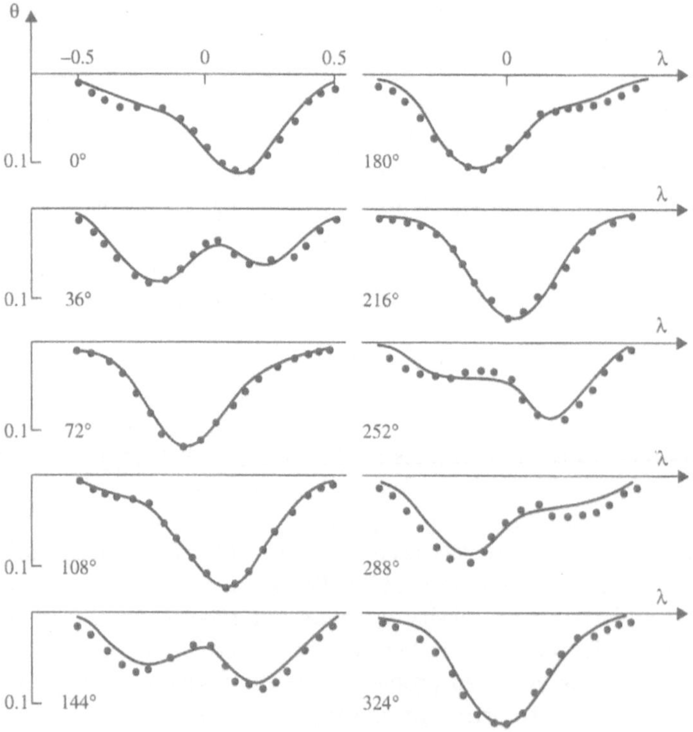

Figure 7.10 The line profiles: model · · · ·; calculated —. The mean square error equals 0.005.

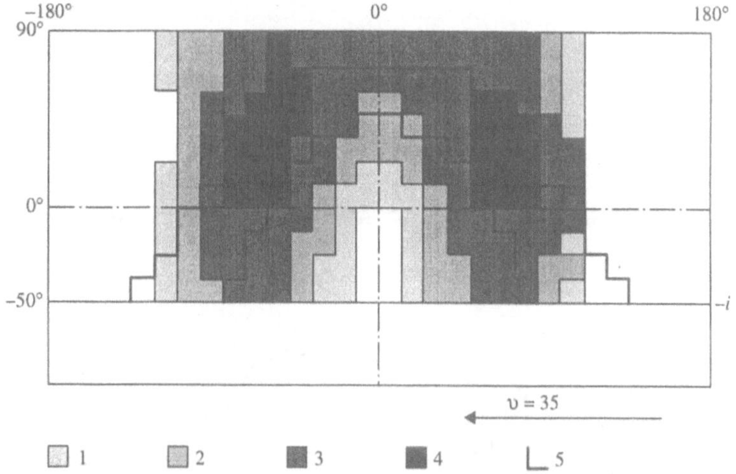

Figure 7.11 The map of the distribution of local equivalent widths (model 2): (1) 0.02–0.05 Å; (2) 0.05–0.08 Å; (3) 0.08–0.11 Å; (4) 0.11–0.14 Å; (5) contours of the initial distribution.

instance, for the line λ 3856 Å Si II the variations of ξ from 0.1 antg to 0.2 antg Si tends to change in abundance approximately in ten times. The results of solving the modelling problem described above illustrate a possibility of splitting the regions with a higher value of the local equivalent width. The spots divided by certain intervals (more than 50°) emerge with sufficient reliability.

We have taken the following values of the needed parameters: $v_{eq} = 35$ km/sec, $i = 50°$, $\varkappa_l = 0.1989 \times 10^{-3}$, $\varkappa_d = 0.0468$, $c_1 = 0.7$, $c_2 = 2.75 \times 10^{-3}$, $\lambda_0 = 3856$ Å and the line width $\Delta\lambda = 1.5$ Å. We used the grid in the wavelength consisting of 19 points. The profiles corresponding to 10 phases uniformly distributed within period were explored. The number of domains where the function preserves a constant value in the process of solving the inverse problem is equal to $11 \times 30 = 330$. In the model distribution the value of the local equivalent width in spots is 0.2 Å. No lines emerge outside the spots. In accordance with the alternative principle the regularization parameter is taken to be $\alpha \sim 10^{-7}$.

Before giving further motivation, we take up the modelling problem related to the distribution of local equivalent widths $\xi(M)$ other than 0 in the regions whose boundary is shown in Fig. 7.11. In that region, τ is a constant equal to 1×10^3. The map of the distribution $\xi(M)$ on the surface obtained by the method described above is depicted in the same figure. Observe that in the initial distribution the generating region is a ring of a large radius. This modelling problem allows to determine the distribution on different regions on the star's surface. In this case, as in the preceding

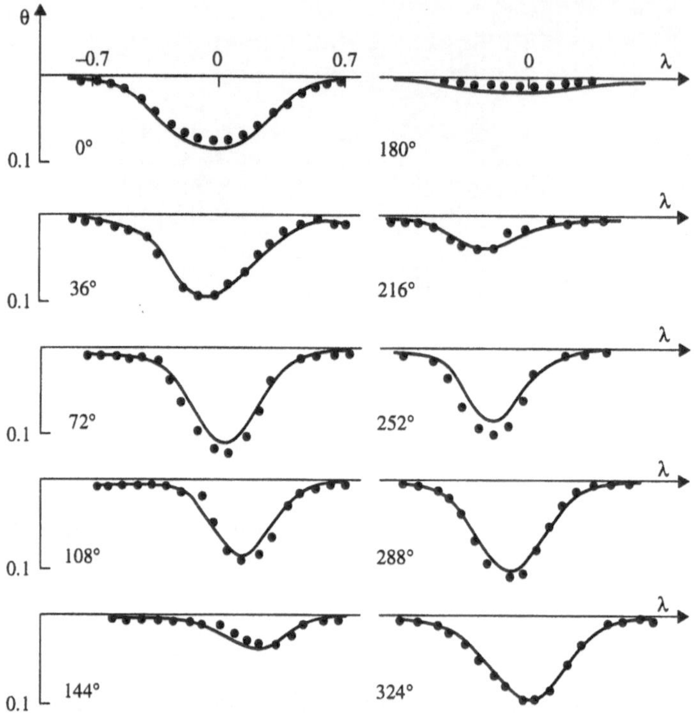

Figure 7.12 The line profiles: model $\cdots\cdot$; calculated —. The mean square error equals 0.008.

one, the line profiles shown in Fig. 7.12 are well-approximated. The value of the regularization parameter equals now $\alpha \sim 10^{-9}$, while the remaining parameters are the same as in the preceding modelling problem.

By means of the algorithm developed in this section we have been considering profiles of the spectral line $\lambda 3862$ ÅSi II in the spectrum of Ap-star CU Vir (HD 124 224) obtained by T.A. Ryabtchikova in the Crimea Astrophysical Laboratory by applying telescopes of 2.6 m with dispersion 4 Å/mm as well as of 125 cm with dispersion 8 Å/mm. To perform calculations we accepted $i = 40°$, $v_{eq} = 210$ km/c, $\varkappa_l = 0.1989 \times 10^{-3}$, $\varkappa_d = 0.468 \times 10^{-1}$, $c_1 = 0.6$, $c_2 = 2.75 \times 10^{-3}$ and the value of the coefficient of the limb darkening was equal to 0.6 (a more detailed description can be found in Goncharskiĭ, Ryabchikova *et al.* (1983) and Goncharskiĭ, Cherepashchuk and Yagola (1985)). For a star with $T_{ef} = 13000$ K we keep $u_2 = 0.5$. The value of the function $\tau(\Phi, l)$ was determined for 160 regions on the surface. In Figs. 7.13–7.14 we present the map of the distribution of local equivalent widths demonstrating the correspondence between the observed and calculated profiles. At each point the error in determining

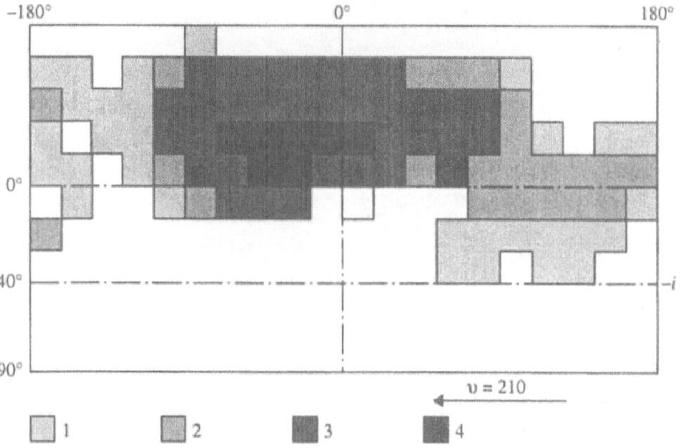

Figure 7.13 HD 124224, λ 3862 Å Si II. The map of the distribution of local equivalent widths: (1) 0.1–0.2 Å; (2) 0.2–0.3 Å; (3) 0.3–0.4 Å; (4) 0.4–0.5 Å.

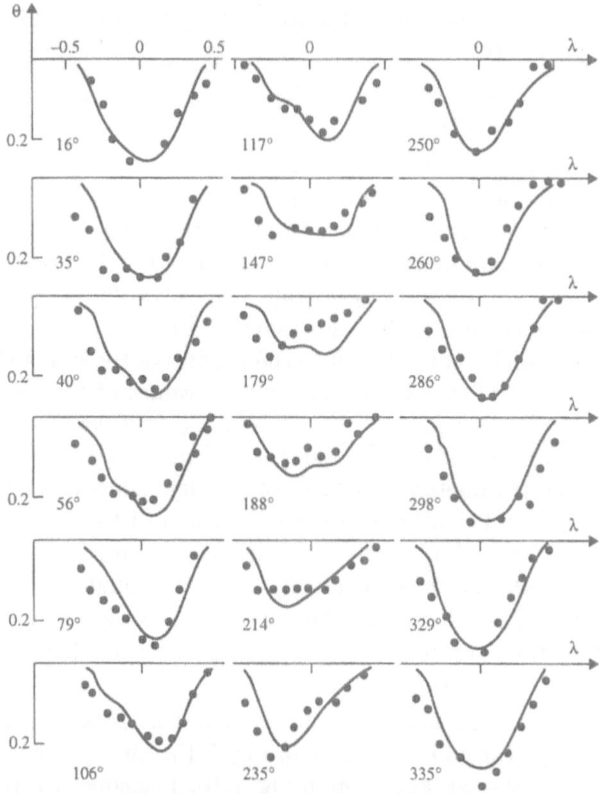

Figure 7.14 HD 124224, λ Å Si II. The line profiles: observed · · ·; calculated —.

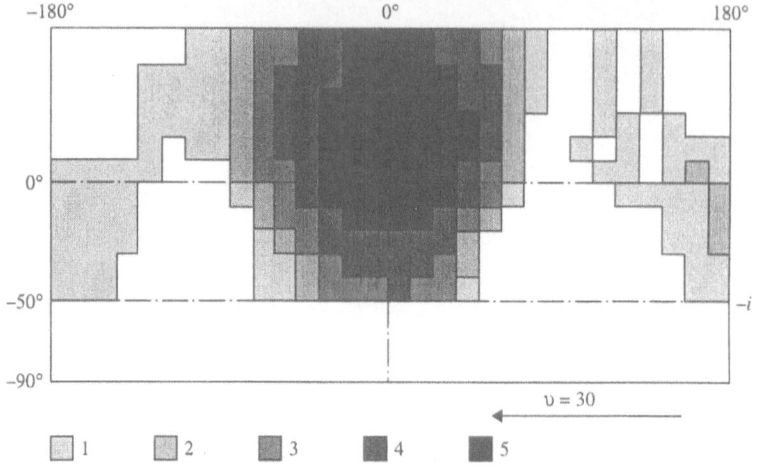

Figure 7.15 HD 112412, λ 4129 Å Eu II. The map of the distribution of local equivalent widths: (1) 0.35–0.70 Å; (2) 0.70–0.105 Å; (3) 0.105–0.140 Å; (4) 0.140–0.175 Å; (5) 0.175–0.210 Å.

experimental profiles is about 0.02. As readily seen from Fig. 7.13, the value of the equivalent line width ranges from 0 to 0.5 Å on the surface, which reflects the excess of silicon up to 90 times as compared with its standard concentration in the star's atmosphere. On CU Vir silicon is only localized in one region with more higher concentration. The value of the regularization parameter was taken to be $\alpha \sim 10^{-9}$.

This method applies equally well to the profiles of the line λ4129 ÅEu II in the spectrum α^2 CVn (HD 112412) whose values of the angle and the equatorial velocity are equal to $i = 50°$ and $v_{eq} = 30$ km/sec. The values of $\varkappa_l$ and $\varkappa_d$ are the same as in the case of the star CU Vir; $u_1 = 0.6$ and $u_2 = 0$. The parameters c_1 and c_2 of the dependency $h(\tau)$ are chosen so as to satisfy a sufficiently accurate restoration of the observed profiles (the reasonings for making such a choice of parameters are justified by Goncharskiĭ, Ryabchikova *et al.* (1983)) and are equal to $c_1 = 0.8$ and $c_2 = 1.25 \times 10^{-2}$. The values of the function $\tau(\Phi, l)$ were determined for 330 regions on the star's surface. Much progress has been achieved by serious developments of the existing algorithms.

The advantage of the new algorithm resulted in obtaining the map of the distribution of europium is shown in Fig. 7.15. The appropriate profiles (calculated and observed) are given in Fig. 7.16. The concentration of Eu is of significant growth in one localized region on the surface at the longitute between -50° and +50°, where it is 10^6 times the standard one. However, at the longitude about 180° there is one more region with a higher concentra-

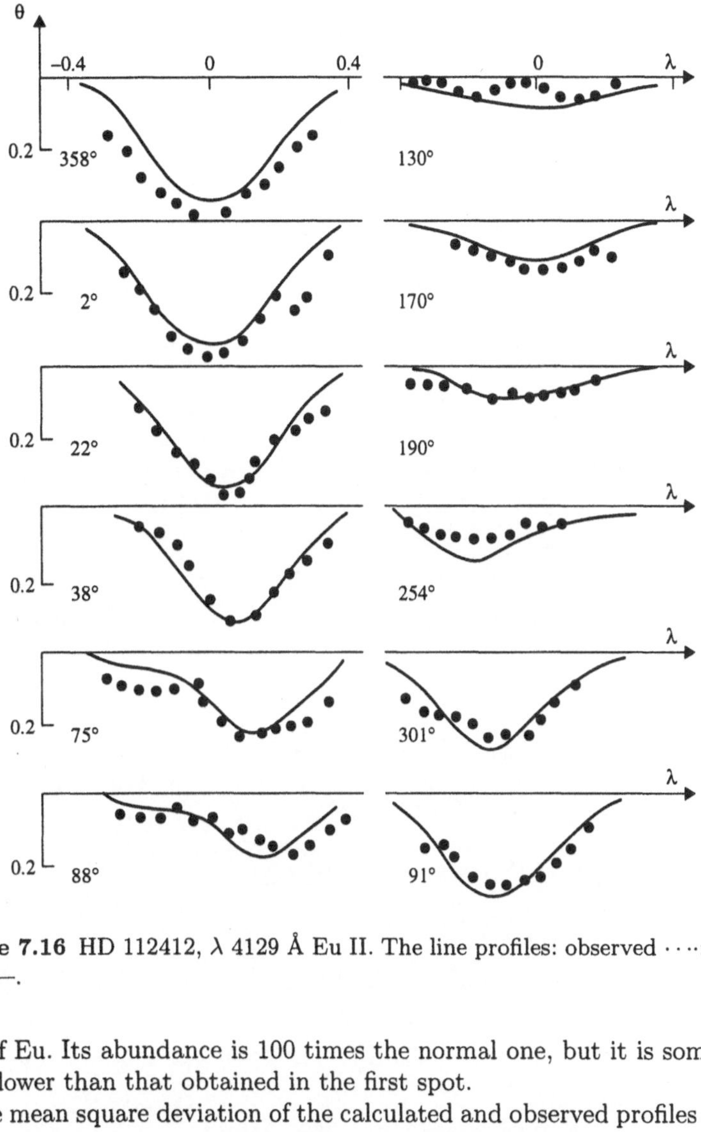

Figure 7.16 HD 112412, λ 4129 Å Eu II. The line profiles: observed ····; calculated —.

tion of Eu. Its abundance is 100 times the normal one, but it is somewhat more lower than that obtained in the first spot.

The mean square deviation of the calculated and observed profiles is 0.04 approximately coinciding with 10% of maximum of the observed profiles; the value of the regularization parameter was taken to be $\alpha \sim 10^{-7}$.

The reader can find other results connected with applications of this method in Goncharskiĭ, Stepanov *et al.* (1982), Goncharskiĭ, Ryabchikova *et al.* (1983), Goncharskiĭ, Cherepashchuk and Yagola (1985).

References

Ageev, A.L. (1980) Regularization of nonlinear operator equations in the class of discontinuous functions. *USSR Comput. Mathem. and Mathem. Phys.*, **20**, 1–9.

Ageev, A.L. and Vasin, V.V. (1979) On convergence of the generalized discrepancy method and its discrete approximations, in *Research in Mathematical Analysis*, Ural's University Press, Sverdlovsk, 3–18 (in Russian).

Albert, A. (1972) *Regression and the Moore-Penrose Pseudoinverse*, Academic Press, New York–London.

Aller, L.H. (1961) *The Abundance of Elements*, Interscience Publ. Inc., New York–London.

Antipin, V.Ya. (1973) On a unified approach to methods for solving ill-posed extremal problems. *Vestnik Moscow University, Ser: Mathem.-Mech.*, **2**, 61–67 (in Russian); English transl. in *Moscow University Mathematics Bulletin.*

Arsenin, V.Ya. (1965) On discontinuous solutions of equations of the first kind. *USSR Comput. Mathem. and Mathem. Phys.*, **5**, 202–209.

Arsenin, V.Ya. (1977) *Methods for Solving Ill-Posed Problems*, Moscow Engineering Physics Institute Publications, Moscow (in Russian).

Arsenin, V.Ya. and Ivanov, V.V. (1968) The solution of certain convolution type integral equations of the first kind by the regularization method. *USSR Comput. Mathem. and Mathem. Phys.*, **8**, 88–106.

Bakhvalov, N.S. (1992) *Numerical Methods*, Kluwer, Dordrecht.

Bakushinskiĭ, A.B. (1965) A numerical method for solving Fredholm integral equations of the first kind. *USSR Comput. Mathem. and Mathem. Phys.*, **5**, 226–233.

Bakushinskiĭ, A.B. (1967) A general method for constructing regularizing algorithms for linear ill-posed equations in Hilbert space. *USSR Comput. Mathem. and Mathem. Phys.*, **7**, 279–287.

Bakushinskiĭ, A.B. (1970) The extension of the discrepancy principle. *USSR Comput. Mathem. and Mathem. Phys.*, **10**, 288–293.

Bakushinskiĭ, A.B. (1979) The principle of iterative regularization. *Zh. Vychisl. Mat. i Mat. Fiz.*, **19**, 1040–1043 (in Russian); English transl. in *USSR Comput. Mathem. and Mathem. Phys.*

Bakushinskiĭ, A.B. (1984) Some remarks on the choice of regularization parameter using quasioptimality and ratio criteria. *USSR Comput. Mathem. and Mathem. Phys.*, **24**, 181–182.

Bakushinskiĭ, A.B. and Goncharskiĭ, A.V. (1989) *Iterative Methods for Solving Ill-Posed Problems*, Nauka, Moscow (in Russian).

Bakushinskiĭ, A.B. and Goncharskiĭ, A.V. (1994) *Ill-Posed Problems: Theory and*

Applications, Kluwer, Dordrecht.

Björck, A. and Eldén, L. (1979) *Methods of Numerical Algebra for Solving Ill-Posed Problems*, Linköping University, Dept. of Mathematics, Preprint LITH–MAT–R–33–1979.

Braun, P.A. and Kiselev, A.A. (1983) *Introduction to the Theory of Molecular Spectrum*, Leningrad University Press, Leningrad (in Russian).

Budak, B.M. and Berkovich, E.M. (1971) On the approximation of extremal problems. Part I. *USSR Comput. Mathem. and Mathem. Phys.*, **11**, 45–66.

Budak, B.M., Vignoli, A. and Gaponenko, Yu.L. (1969) On a certain regularization of extremal problems in the case of a continuous convex functional. *Dokl. Akad. Nauk SSSR*, **184**, 12–15 (in Russian); English transl. in *Soviet Mathem. Dokl.*

Buša, J. (1987) On a regularization method for solving systems of linear algebraic equations. *Dokl. Akad. Nauk SSSR*, **295**, 11–14 (in Russian); English transl. in *Soviet Mathem. Dokl.*

Clarkson, J.A. and Adams, C.R. (1933) On definition of bounded variation for functions of two variables. Trans. Amer. Mathem. Soc., **35**, 824–854.

Diestel, J. (1975) *Geometry of Banach Spaces – Selected Topics. Lect. Notes in Mathematics*, Springer-Verlag, Berlin–Heidelberg, **485**, N XI.

Dmitriev, M.G. and Poleshchuk, V.S. (1972) On the regularization of a certain class of unstable extremal problems. *USSR Comput. Mathem. and Mathem. Phys.*, **12**, 289–293.

Dunford, N. and Schwartz, J.T. (1971) *Linear Operators*, Volume 1 (2nd edition), Wiley, New York.

Eremin, Yu.A. and Leonov, A.S. (1975) The calculation of stable difference schemes for second-order alternating linear differential operators. *USSR Comput. Mathem. and Mathem. Phys.*, **15**, 86–94.

Fiacco, A. and McCormick, G. (1968) *Nonlinear Programming: Sequential Unconstrained Minimization Techniques*, Wiley, New York.

Forsythe, G., Malcolm, M. and Moler, C. (1977) *Computer Methods for Mathematical Computations*, Prentice-Hall, Englewood Cliffs, NJ.

Gantmakher, F.R. (1967) *The Theory of Matrices* (3rd edn), Nauka, Moscow (in Russian).

Gaponenko, Yu.L. (1978) A regularizer in the space of continuous functions. *USSR Comput. Mathem. and Mathem. Phys.*,**18**, 83–87.

Gaponenko, Yu.L. (1980) The method of sequential approximation for solving nonlinear extremal problems. *Izv. Vuz., Ser: Matem.*, **5**, 12–16 (in Russian); English transl. in *Soviet Mathematics*, Allerton Press.

Gaponenko, Yu.L. (1985) The accuracy of the solution of a nonlinear ill-posed problem at a finite error level. *USSR Comput. Mathem. and Mathem. Phys.*, **25**, 81–85.

Gaponenko, Yu.L. (1989) *Ill-Posed Problems on Weak Compacts*, Moscow University Press, Moscow (in Russian).

Gavurin, M.K. (1971) *Lectures on Numerical Methods*, Nauka, Moscow (in Russian).

Gilyazov, S.F. (1987) *Methods for Solving Linear Ill-Posed Problems*, Moscow University Press, Moscow (in Russian).

Gilyazov, S.F. and Morozov, V.A. (1984) The optimal regularization of ill-posed normally solvable operator equations. *USSR Comput. Mathem. and Mathem. Phys.*, **24**, 89–92.

Glasko, V.B., Gushchin, G.V. and Starostenko, V.I. (1976) On the application of the Tikhonov regularization to solving nonlinear systems of equations. *USSR Comput. Mathem. and Mathem. Phys.*, **16**, 1–9.

Golub, G.H. (1968) Least squares, singular values and matrix approximation. *Aplikace Matematiky*, **13**, 44–51.

Goncharskiĭ, A.V. and Stepanov, V.V. (1979) On a uniform approximation of the solution of bounded variation to ill-posed problems. *Dokl. Akad. Nauk SSSR*, **248**, 20–22 (in Russian); English transl. in *Soviet Mathem. Dokl.*

Goncharskiĭ, A.V. and Stepanov, V.V. (1980) Numerical methods for solving ill-posed problems on compact sets. *Vestnik Moscow University, Ser: Vychisl. Matem. i Kibern.*, **15**, 12–18 (in Russian); English transl. in *Moscow University Computational Mathematics and Cybernetics.*

Goncharskiĭ, A.V. and Yagola, A.G. (1969) On a uniform approximation of the monotone solution of ill-posed problems. *Dokl. Akad. Nauk SSSR*, **184**, 771–773 (in Russian); English transl. in *Soviet Mathem. Dokl.*

Goncharskiĭ, A.V., Cherepashchuk, A.M. and Yagola, A.G. (1978) *Numerical Methods for Solving Inverse Problems in Astrophysics*, Nauka, Moscow (in Russian).

Goncharskiĭ, A.V., Cherepashchuk, A.M. and Yagola, A.G. (1985) *Ill-Posed Problems in Astrophysics*, Nauka, Moscow (in Russian).

Goncharskiĭ, A.V., Gushchina, L.G., Leonov, A.S. and Yagola, A.G. (1972) Some algorithms for finding approximate solutions of ill-problems in a class of monotone functions. *USSR Comput. Mathem. and Mathem. Phys.*, **12**, 1–18.

Goncharskiĭ, A.V., Leonov, A.S. and Yagola, A.G. (1971) The solution of two-dimensional Fredholm equations of the first kind with a kernel depending on the difference of arguments. *USSR Comput. Mathem. and Mathem. Phys.*, **11**, 253–260.

Goncharskiĭ, A.V., Leonov, A.S. and Yagola, A.G. (1972a) Some generalization of the discrepancy principle for an operator specified with error. *Dokl. Akad. Nauk SSSR*, **203**, 1238–1239 (in Russian); English transl. in *Soviet Mathem. Dokl.*

Goncharskiĭ, A.V., Leonov, A.S. and Yagola, A.G. (1972b) A regularizing algorithm for solving ill-posed problems with approximately specified operator. *USSR Comput. Mathem. and Mathem. Phys.*, **12**, 286–290.

Goncharskiĭ, A.V., Leonov, A.S. and Yagola; A.G. (1973) The generalized discrepancy principle. *USSR Comput. Mathem. and Mathem. Phys.*, **13**, 25–37.

Goncharskiĭ, A.V., Leonov, A.S. and Yagola, A.G. (1974a) Finite-difference approximation of linear ill-posed problems. *USSR Comput. Mathem. and Mathem. Phys.*, **14**, 14–23.

Goncharskiĭ, A.V., Leonov, A.S. and Yagola, A.G. (1974b) On the discrepancy principle for solving nonlinear ill-posed problems. *Dokl. Akad. Nauk SSSR*, **15**, 166–168 (in Russian); English transl. in *Soviet Mathem. Dokl.*

Goncharskiĭ, A.V., Leonov, A.S. and Yagola, A.G. (1974c) On the solution of ill-posed problems with approximately specified operator, in *Proceedings of the all-union school of young mathematicians "Methods for Solving Ill-Posed*

Problems and Their Applications", Moscow University Press, Moscow, 39–43 (in Russian).

Goncharskiĭ, A.V., Leonov, A.S. and Yagola, A.G. (1975) Applicability of the discrepancy principle in the case of nonlinear ill-posed problems and a new regularizing algorithm for solving them. *USSR Comput. Mathem. and Mathem. Phys.*, **15**, 8–16.

Goncharskiĭ, A.V., Ryabchikova, T.A., Stepanov, V.V., Khokhlova, V.L. and Yagola, A.G. (1983) On maps of chemical elements at the surface of Ap-stars. *Astron. J.*, **60**, 83–90 (in Russian).

Goncharskiĭ, A.V., Stepanov, V.V., Khokhlova, V.L. and Yagola, A.G. (1977) Restoration of local profiles of lines from observations in the spectrum of Ap-stars, *Letters to Astron. J.*, **3**, 278–282 (in Russian).

Goncharskiĭ, A.V., Stepanov, V.V., Khokhlova,V.L. and Yagola, A.G. (1982) On maps of chemical elements at the surface of Ap-stars. *Astron. J.*, **59**, 1146–1156 (in Russian).

Greville, T.N.E. (1960) Some applications of the pseudoinverse of a matrix. *SIAM Review*, **2**, 15–23.

Hansen, P.C. (1987) The truncated SVD as a method for regularization. *BIT*, **27**, 534–553.

Hansen, P.C. (1992) Numerical tools for analysis and solution of Fredholm integral equations of the first kind. *Inverse Problems*, **8**, 849–872.

Himmelblau, D. (1971) *Applied Nonlinear Programming*, McGraw Hill, New York.

Ioffe, V.V., Kostikov, R.R. and Razin, V.V. (1984) *Physical Methods for the Determination of Structure of Organic Compounds*, Vischa Shkola, Moscow (in Russian).

Ivanov, V.K. (1962a) Integral equations of the first kind and the approximate solution of an inverse potential problem. *Dokl. Akad. Nauk SSSR*, **142**, 997–1000 (in Russian); English transl. in *Soviet Mathem. Dokl.*

Ivanov, V.K. (1962b) On linear ill-posed problems. *Dokl. Akad. Nauk SSSR*, **145**, 270–272 (in Russian); English transl. in *Soviet Mathem. Dokl.*

Ivanov, V.K. (1966) The approximate solution of operator equations of the first kind. *USSR Comput. Mathem. and Mathem. Phys.*, **6**, 197–205.

Ivanov, V.K. (1969) Ill-posed problems in topological spaces. *Sibirian Mathem. J.*, **10**, 1065–1074.

Ivanov, V.K., Vasin, V.V. and Tanana, V.P. (1978) *The Theory of Linear Ill-Posed Problems and Its Applications*, Nauka, Moscow (in Russian).

Kantorovich, L.V. and Akilov, G.P. (1977) *Functional Analysis*, Nauka, Moscow (in Russian).

Karmanov, V.G. (1975) *Mathematical Programming*, Nauka, Moscow (in Russian).

Kasparov, K.N. and Leonov, A.S. (1975) On the photoemission method of spectrum measurements and a new way of their mathematical treatment. *Appl. Spectroscop. J.*, **22**, 491–498 (in Russian).

Khokhlova, V.L. (1976) On the formalization of the inverse problem of reconstructing local profiles from observations in spectrum of Ap-stars. *Astron. Nachr.*, **297**, 203–206.

Kochetov, I.I. (1976) A new way of choosing the regularization parameter. *USSR*

Comput. Mathem. and Mathem. Phys., **16**, 215–218.

Kochikov, I.V. and Kuramshina, G.M. (1985) A package of programs for calculating force fields of polyatomic molecules by means of the Tikhonov regularization method. *Vestnik Moscow University, Ser: Khimiya.*, **26**, 354–358 (in Russian); English transl. in *Moscow University Chemistry Bulletin.*

Kochikov, I.V., Kuramshina, G.M. and Yagola, A.G. (1987) Stable numerical methods for solving certain problems of vibrational spectroscopy. *USSR Comput. Mathem. and Mathem. Phys.*, **27**, 33–39.

Kochikov, I.V., Matvienko, A.N. and Yagola, A.G. (1983a) On the solution of incompatible operator equations by means of the generalized discrepancy principle, in *The Theory and Methods for Solving Ill-Posed Problems*, Novosibirsk University Press, Novosibirsk, 43–48 (in Russian).

Kochikov, I.V., Matvienko, A.N. and Yagola, A.G. (1983b) The generalized discrepancy principle for solving incompatible equations. *USSR Comput. Mathem. and Mathem. Phys.*, **23**, 78–80.

Kochikov, I.V., Matvienko, A.N. and Yagola, A.G. (1984) A modification of the generalized discrepancy principle. *USSR Comput. Mathem. and Mathem. Phys.*, **24**, 1087–1090.

Kochikov, I.V., Matvienko, A.N. and Yagola, A.G. (1987) A regularization method for solving incompatible nonlinear operator equations. *USSR Comput. Mathem. and Mathem. Phys.*, **27**, 91–92.

Kochikov, I.V., Kuramshina, G.M., Pentin, Yu.A. and Yagola, A.G. (1981) Regularizing algorithm for solving the inverse vibrational problem. *Dokl. Akad. Nauk SSSR*, **261**, 1104–1106 (in Russian); English transl. in *Soviet Mathem. Dokl.*

Kochikov, I.V., Kuramshina, G.M., Pentin, Yu.A. and Yagola, A.G. (1983) On the stable solution of the inverse vibrational problem, in *The Theory and Methods for Solving Ill-Posed Problems*, Sibirian Computer Center Publications, Novosibirsk, 124–125 (in Russian).

Kochikov, I.V., Kuramshina, G.M., Pentin, Yu.A. and Yagola, A.G. (1984) A stable method for force field calculations in redundant systems of internal coordinates. *Teor. Eksper. Khimiya*, **20**, 69–75 (in Russian).

Kochikov, I.V., Kuramshina, G.M., Chernik, S.I. and Yagola, A.G. (1986) An algorithm for determining the normal solution of the inverse vibrational problem based on Monte-Carlo method. *Vestnik Moscow University, Ser: Khimiya*, **27**, 597–601 (in Russian); English transl. in *Moscow University Chemistry Bulletin.*

Kochikov, I.V., Yagola, A.G., Kuramshina, G.M., Kovba, V.M. and Pentin, Yu.A. (1984) Force fields and mean amplitudes of vibration of chromium, molybdenum and tungsten oxotetrafluorides. *J. Molec. Structure*, **106**, 355–360.

Kochikov, I.V., Yagola, A.G., Kuramshina, G.M., Kovba, V.M. and Pentin, Yu.A. (1985) Calculation of force fields of chromium, molybdenum and tungsten hexafluorides and dioxodifluorides by means of the Tikhonov regularization method. *Spectrochim Acta, Ser: A*, **41**, 185–189.

Kolmogorov, A.N. and Fomin, S.V. (1968) *Elements of the Theory of Functions and Functional Analysis*, Nauka, Moscow (in Russian).

Koptev, G.S. and Pentin, Yu.A. (1977) *The Calculation of Molecular Vibrations*, Moscow University Press, Moscow (in Russian).

Korotaev, A.L., Kuznetsov, P.I. , Leonov, A.S. and Rodoman, V.E. (1986) On the determination of optimal regime for injecting antibactrim. *Autom. Remove Control*, **1**, 100–106.

Krekoten', S.P., Leonov, A.S. and Sukharevskiĭ, V.G. (1986) On optimal mathematical design of electromagnetic systems. *Dokl. Akad. Nauk SSSR*, **287**, 312–316 (in Russian); English transl. in *Soviet Mathem. Dokl.*

Kryanev, A.V. and Tsupko–Sitnikov, M.V. (1988) Nonlinear statistical methods for solving ill-conditioned systems of linear equations and their application to experimental data processing, in *Conditional-Posed Problems in Mathematical Physics and Analysis*, Krasnoyarsk University Press, Krasnoyarsk, 185–180 (in Russian).

Lancaster, P. and Tismensky, M. (1985) *The Theory of Matrices*, Academic Press, New York.

Lánczos, C. (1956) *Applied Analysis*, Prentice-Hall, Englewood Cliffs, NJ.

Lawson, C.L. and Hanson, R.J. (1974) *Solving Least Squares Problems*, Prentice-Hall, Englewood Cliffs, NJ.

Leonov, A.S. (1975) On the construction of stable difference schemes for solving nonlinear boundary value problems. *Dokl. Akad. Nauk SSSR*, **224**, 525–528 (in Russian); English transl. in *Soviet Mathem. Dokl.*

Leonov, A.S. (1976) Some aspects of realization of the regularizing algorithm of the generalized discrepancy, in *Implementation and Interpretation of Physical Experiment*, Moscow University Press, Moscow, **4**, 69–81 (in Russian).

Leonov, A.S. (1978) On the choice of regularization parameter by means of the quasioptimality and ratio criteria. Zh. Vychisl. Mat. i Mat. Fiz., **18**, 1363–1376 (in Russian); English transl. in *USSR Comput. Mathem. and Mathem. Phys.*

Leonov, A.S. (1979a) On algorithms for the approximate solution of nonlinear ill-posed problems with perturbed operator. *Soviet Mathem. Dokl.*, **20**, 301–305.

Leonov, A.S. (1979b) On the choice of regularization parameter for solving nonlinear ill-posed problems with approximately specified operator. Zh. Vychisl. Mat. i Mat. Fiz., **19**, 1363–1376 (in Russian); English transl. in *USSR Comput. Mathem. and Mathem. Phys.*

Leonov, A.S. (1980) On the regularization of ill-posed problems with discontinuous solutions and the application of this methodology to solving some nonlinear equations. *Soviet Mathem. Dokl.*, **21**, 25–28.

Leonov, A.S. (1982a) Piecewise-uniform regularization of ill-posed problems with discontinuous solutions. *USSR Comput. Mathem. and Mathem. Phys.*, **22**, 20–36.

Leonov, A.S. (1982b) On the connection between the generalized discrepancy method and the generalized discrepancy principle for solving nonlinear ill-posed problems. *USSR Comput. Mathem. and Mathem. Phys.*, **22**, 13–21.

Leonov, A.S. (1982c) On the application of the generalized discrepancy principle to solving ill-posed extremal problems. *Soviet Mathem. Dokl.*, **25**, 227–231.

Leonov, A.S. (1982d) On the criteria of choosing the regularization parameter while solving ill-posed problems, in *Methods for Solving Ill-Posed Problems and Their Applications*, Sibirian Computer Center Publications, Novosibirsk, 77–84 (in Russian).

Leonov, A.S. (1983a) The minimal pseudoinverse matrix method and the solution of ill-posed problems of linear algebra on its basis, in *The Theory and Methods for Solving Ill-Posed Problems and Their Applications*, Sibirian Computer Center Publications, Novosibirsk, 49–52 (in Russian).

Leonov, A.S. (1983b) Applying regularizing algorithms to solving some problems of medical radiology, in *The Theory and Methods for Solving Ill-Posed Problems and Their Applications*, Sibirian Computer Center Publications, Novosibirsk, 132 (in Russian).

Leonov, A.S. (1983c) Solving linear ill-posed problems on the basis of the modified quasioptimality criterion, *Matem. Sbornik*, **122**, 405–415 (in Russian).

Leonov, A.S. (1985a) The minimal pseudoinverse matrix method for solving ill-posed problems of linear algebra. *Dokl. Akad. Nauk SSSR*, **285**, 36–40 (in Russian); English transl. in *Soviet Mathem. Dokl.*

Leonov, A.S. (1985b) The approximate calculation of a pseudoinverse matrix by means of the generalized discrepancy principle. *USSR Comput. Mathem. and Mathem. Phys.*, **25**, 181–182.

Leonov, A.S. (1986) On algorithms for solving ill-posed extremal problems. *Matem. Sbornik*, **129**, 218–231 (in Russian).

Leonov, A.S. (1987a) The minimal pseudoinverse matrix method. *USSR Comput. Mathem. and Mathem. Phys.*, **27**, 107–117.

Leonov, A.S. (1987b) The numerical realization of piecewise-uniformly regularizing algorithms. *USSR Comput. Mathem. and Mathem. Phys.*, **27**, 88–91.

Leonov, A.S. (1988a) Optimality in accuracy order of the generalized discrepancy principle and several other algorithms for solving nonlinear ill-posed problems with inexact data. *Sibirian Mathem. J.*, **29**, 85–94.

Leonov, A.S. (1988b) Numerical algorithms with nonapriori choice of regularization parameter for solving nonlinear ill-posed problems, in *Proceeding of all-union seminar "Inverse Problems and Identification of Heat Exchange Processes"*, Moscow Aviation Institute Publications, Moscow, 6–7 (in Russian).

Leonov, A.S. (1988c) Numerical algorithms for solving nonlinear ill-posed problems with nonapriori choice of regularization parameter. VINITI Publications, Moscow, 346-B-88 (in Russian).

Leonov, A.S. (1988d) Optimality in accuracy order of regularizing algorithms for solving nonlinear operator equations, in *Condional-Posed Problems in Mathematical Physics and Analysis*, Krasnoyarsk University Press, Krasnoyarsk, 118–124 (in Russian).

Leonov, A.S. (1990) Optimality in accuracy order of some algorithms for solving ill-posed extremal problems. *Izv. Vuz., Ser: Matem.*, **6**, 30–38 (in Russian); English transl. in *Soviet Mathematics*, Allerton Press.

Leonov, A.S. (1991) On the theory of the minimal pseudoinverse matrix method. *Soviet Mathem. Dokl.*, **42**, 327–332.

Leonov, A.S. (1992) On the accuracy of Tikhonov regularizing algorithms and the quasioptimal choice of regularization parameter. *Soviet Mathem. Dokl.*, **44**, 711–716.

Leonov, A.S. and Rusin, S.P. (1988) Optimal design of the radiation concentrators in cosmic technology, in *Proceedings of XXIII Tsiolkovskiĭ conference: Problems of Cosmic Production*, Moscow, 109–113 (in Russian).

Leonov, A.S. and Suleĭmanova, M.S. (1985) The data processing in dynamical radionuclid studies on the basis of methods for solving ill-posed problems. *Medical radiology*, **12**, 60–63 (in Russian).

Liskovets, O.A. (1973) The method of ε-quasisolutions for equations of the first kind. *Diff. Uravneniya*, **9**, 1851–1861 (in Russian); English transl. in *Differ. Equations.*

Liskovets, O.A. (1981) *Variational Methods for Solving Unstable Problems*, Nauka i Tekhnika, Minsk (in Russian).

Lyusternik, L.A. and Sobolev, V.I. (1982) *A Concise Course on Functional Analysis*, Vischa Shkola, Moscow (in Russian).

Maindonald, J.H. (1984) *Statistical Computation*, Wiley, New York.

Matveev, V.S. (1972) The approximate representation of the absorption coefficient and equivalent widths of lines with Foyght's contour. *Appl. Spectroscop. J.*, **16**, 228–233 (in Russian).

Meleshko, V.I. and Zadachin, V.M. (1987) Factorization and pseudoinversion of singular perturbed matrices with alternating signs. *Izv. Vuz., Ser: Matem.*, **11**, 42–50 (in Russian); English transl. in *Soviet Mathematics*, Allerton Press.

Mil'man, V.D. (1971) Geometric theory in Banach spaces. Part II. Geometry of unit sphere. *Uspekhi Matem. Nauk*, **26**, 73–149 (in Russian) English transl. in *Mathem. Surveys.*

Moore, E.N. (1920) On the reciprocal of the general algebraic matrix, *Bull. Amer. Mathem. Soc.*, **26**, 394–395.

Morozov, V.A. (1966a) On the solution of functional equations by the regularization method. *Dokl. Akad. Nauk SSSR*, **167**, 510–512 (in Russian); English transl. in *Soviet Mathem. Dokl.*

Morozov, V.A. (1966b) The regularization of ill-posed problems and the choice of the regularization parameter. *USSR Comput. Mathem. and Mathem. Phys.*, **6**, 242–251.

Morozov, V.A. (1967) *The Regularization Method and Its Applications.* Ph. D. Thesis, Moscow University Press (in Russian).

Morozov, V.A. (1968) The discrepancy principle for solving operator equations by the regularization method. *USSR Comput. Mathem. and Mathem. Phys.*, **8**, 663–887.

Morozov, V.A. (1969a) On the regularization of certain classes of extremal problems, in *Numerical Methods and Programming*, Moscow University Press, Moscow, **12**, 24–37 (in Russian).

Morozov, V.A. (1969b) Pseudosolutions. *USSR Comput. Mathem. and Mathem. Phys.*, **9**, 196–203.

Morozov, V.A. (1973a) On a new approach to solving first-order linear equations with approximately specified operators, in *Proceedings of the first conference of young scientists of Computer Science Department of Moscow State University*, Moscow University Press, Moscow, 22–28 (in Russian).

Morozov, V.A. (1973b) On general conditions for the regularization of ill-posed variational problems, in *Proceedings of the first conference of young scientists of Computer Science Department of Moscow State University*, Moscow University Press, Moscow, 140–164 (in Russian).

Morozov, V.A. (1973c) Linear and nonlinear ill-posed problems. *Itogi Nauki i Tekhniki, Ser: Matem. Anal.*, **11**, 129–178 (in Russian).

Morozov, V.A. (1973d) On the calculation of lower bounds of functionals based on the approximate information. *USSR Comput. Mathem. and Mathem. Phys.*, **13**, 1045–1048.

Morozov, V.A. (1973e) The discrepancy principle for solving incompatible equations by means of Tikhonov regularization. *USSR Comput. Mathem. and Mathem. Phys.*, **13**, 1–16.

Morozov, V.A. (1974a) *Regular Methods for Solving Ill-Posed Problems*, Moscow University Press, Moscow (in Russian).

Morozov, V.A. (1974b) An optimality principle for the error when approximately solving equations with nonlinear operator. *USSR Comput. Mathem. and Mathem. Phys.*, **14**, 1–8.

Morozov, V.A. (1978) Regular methods for solving nonlinear operator equations. *Izv. Vuz., Ser: Matem.*, **11**, 74–86 (in Russian); English transl. in *Soviet Mathematics*, Allerton Press.

Morozov, V.A. (1984) *Regularization Methods for Solving Ill-Posed Problems*, Springer, Berlin–Heidelberg.

Morozov, V.A. and Gilyazov, S.F. (1979) On the numerical solution of certain types of incompatible equations, in *Numerical Analysis in Fortran. Numerical Methods and Instrumental Systems*, Moscow University Press, Moscow, 69–79 (in Russian).

Mudrov, V.I. and Kushko, V.L. (1983) *Methods of Data Processing*, Radio i Svyaz', Moscow (in Russian).

Muzylev, N.V. (1975) An algorithm of a simplified regularization. *USSR Comput. Mathem. and Mathem. Phys.*, **15**, 218–222.

Natanson, I.P. (1974) *The Theory of Functions of the Real Variable* (3rd edn), Nauka, Moscow (in Russian).

Pavlova, V.M. and Khokhlova, V.L. (1980) On the connection between parameters of the analytic representation for the line profile in the star's spectrum. *Sci. Inform. of Astronomy Society of AN SSSR*, **43**, 49–54 (in Russian).

Penrose, R.A. (1955) A generalized inverse of matrices. *Proc. Cambridge Phil. Soc.*, **51**, 406–413.

Phillips, D.L. (1962) A technique for the numerical solution of certain integral equations of the first kind. *J. Assoc. Comput. Mathem.*, **9**, 84–97.

Polak, E. (1971) *Computational Methods in Optimization*, Academic Press, New York–London.

Polyak, B.T. (1974) Optimization methods under constraints, in *Mathematical Analysis* (*Itogi Nauki i Tekhniki*), VINITI Publications, **12**, 147–198 (in Russian).

Pshenichny, B.N. (1993) *Linearization Method*, World Scientific, Singapore.

Pshenichny, B.N. and Danilin, Yu.M. (1975) *Numerical Methods for Extremal Problems*, Nauka, Moscow (in Russian).

Rusin, S.P. and Leonov, A.S. (1987) On optimal mathematical design of high temperature radiators. *Izv. AN SSSR, Ser: Energ. i Transp.*, **4**, 154–158 (in Russian).

Rusin, S.P. and Leonov, A.S. (1991) Applying the regularization method to solving inverse problems of radiation heat exchange. *Izv. AN SSSR, Ser: Energ. i*

Transp., **2**, 142–147 (in Russian).

Rusin, S.P. and Peletskiĭ, V.E. (1987) *Heat Radiation of Cavities*, Energoatomizdat, Moscow (in Russian).

Samarskiĭ, A.A. (1989) *The Theory of Difference Schemes* (5th edn), Nauka, Moscow (in Russian).

Strakhov, V.N. (1970) On the solution of linear ill-posed problems in Hilbert space. *Diff. Uravneniya*, **8**, 1490–1495 (in Russian); English transl. in *Differ. Equations.*

Strakhov, V.N. (1973) Constructing optimal in order approximate solutions of linear conditional-posed problems. *Diff. Uravneniya*, **9**, 1862–1874 (in Russian); English transl. in *Differ. Equations.*

Strakhov, V.N. (1988) On smoothing and transformation of potential fields values by means of the method of "fractional" regularization. *Izv. Akad. Nauk SSSR. Ser: Fiz. Zemli*, **1**, 66–81 (in Russian).

Strakhov, V.N. and Valyashko, G.M. (1979) Adaptive regularization of linear ill-posed problems, in *Proceedings of all-union conference on ill-posed problems*, Ilim, Phrunze, 109–110 (in Russian).

Tanana, V.P. (1975a) On a projection iterative algorithm for operator equations of the first kind with perturbed operator. *Dokl. Akad. Nauk SSSR*, **224**, 1028–1029 (in Russian); English transl. in *Soviet Mathem. Dokl.*

Tanana, V.P. (1975b) Projection methods and finite-dimensional approximation of linear ill-posed problems. *Sibirian Mathem. J.*, **16**, 1301–1307.

Tanana, V.P. (1976) Optimal order methods for solving nonlinear ill-posed problems. *USSR Comput. Mathem. and Mathem. Phys.*, **16**, 219–225.

Tanana, V.P. (1979) On the choice of regularization parameter while solving ill-posed problems, in *Proceedings of all-union conference on ill-posed problems*, Ilim, Phrunze, 113–114 (in Russian).

Tanana, V.P. (1981) *Methods for Solving Operator Equations*, Nauka, Moscow (in Russian).

Tanana, V.P., Rekant, M.A. and Yanchenko, S.I. (1987) *Optimization of Methods for Solving Operator Equations*, Ural's University Press, Sverdlovsk (in Russian).

Tikhonov, A.N. (1943) On stability of inverse problems. *Dokl. Akad. Nauk SSSR*, **39**, 195–198 (in Russian).

Tikhonov, A.N. (1963a) On the solution of ill-posed problems and the regularization method. *Dokl. Akad. Nauk SSSR*, **151**, 501–504 (in Russian); English transl. in *Soviet Mathem. Dokl.*

Tikhonov, A.N. (1963b) On the regularization of ill-posed problems. *Dokl. Akad. Nauk SSSR*, **153**, 49–52 (in Russian); English transl. in *Soviet Mathem. Dokl.*

Tikhonov, A.N. (1964) The solution of nonlinear integral equations. *Dokl. Akad. Nauk SSSR*, **156**, 1296–1299 (in Russian); English transl. in *Soviet Mathem. Dokl.*

Tikhonov, A.N. (1965a) On nonlinear equations of the first kind. *Dokl. Akad. Nauk SSSR*, **161**, 1023–1026 (in Russian); English transl. in *Soviet Mathem. Dokl.*

Tikhonov, A.N. (1965b) On ill-posed problems of linear algebra and stable methods for their solution. *Dokl. Akad. Nauk SSSR*, **163**, 591–594 (in Russian);

English transl. in *Soviet Mathem. Dokl.*

Tikhonov, A.N. (1965c) On the regularization methods for problems of optimal control. *Dokl. Akad. Nauk SSSR*, **162**, 763–765 (in Russian); English transl. in *Soviet Mathem. Dokl.*

Tikhonov, A.N. (1965d) On ill-posed problems of optimal planning and stable methods for solving them. *Dokl. Akad. Nauk SSSR*, **164**, 507–510 (in Russian); English transl. in *Soviet Mathem. Dokl.*

Tikhonov, A.N. (1965e) The stability of algorithms for solving degenerate systems of linear algebraic equations. *USSR Comput. Mathem. and Mathem. Phys.*, **5**, 181–188.

Tikhonov, A.N. (1966a) Ill-posed problems of optimal programming. *USSR Comput. Mathem. and Mathem. Phys.*, **6**, 114–127.

Tikhonov, A.N. (1966b) On stability of the functional optimization problems. *USSR Comput. Mathem. and Mathem. Phys.*, **6**, 28–33.

Tikhonov, A.N. (1980) On approximate systems of linear algebraic equations. *USSR Comput. Mathem. and Mathem. Phys.*, **20**, 1371–1383.

Tikhonov, A.N. (1985) On problems with inaccurate initial information. *Dokl. Akad. Nauk SSSR*, **280**, 559–562 (in Russian); English transl. in *Soviet Mathem. Dokl.*

Tikhonov, A.N. and Arsenin, V.Ya. (1977) *Solution of Ill-Posed Problems*, Wiley, New York.

Tikhonov, A.N. and Glasko, V.B. (1964) The approximate solution of Fredholm integral equations of the first kind. *USSR Comput. Mathem. and Mathem. Phys.*, **4**, 236–247.

Tikhonov, A.N. and Glasko, V.B. (1965) Applying the method of regularization to nonlinear problems. *USSR Comput. Mathem. and Mathem. Phys.*, **5**, 93–107.

Tikhonov, A.N. and Ufimtsev, M.V. (1988) *Statistical Approach to Experimental Data Processing*, Moscow University Press, Moscow (in Russian).

Tikhonov, A.N. and Vasilyev, F.P. (1978) Methods for solving ill-posed expremal problems, in *Mathematical Models and Numerical Methods*, Banach Mathem. Center Publications, Warsaw, **3**, 297–342.

Tikhonov, A.N., Dmitriev, V.I., Chechkin, A.V. and Berezina, N.I. (1973) Development of variational methods for solving problems of antenna design, in *Numerical Methods and Programming*, Moscow University Press, Moscow, **20**, 246–257 (in Russian).

Tikhonov, A.N., Glasko, V.B. and Kriksin, Yu.A. (1979) The question of the quasioptimal choice of regularized approximation. *Dokl. Akad. Nauk SSSR*, **248**, 531–534 (in Russian); English transl. in *Soviet Mathem. Dokl.*

Tikhonov, A.N., Goncharskiĭ, A.V., Stepanov, V.V., and Yagola, A.G. (1983a) *Regularizing Algorithms and Apriori Information*, Nauka, Moscow (in Russian).

Tikhonov, A.N., Goncharskiĭ, A.V., Stepanov, V.V., and Yagola, A.G. (1995) *Numerical Methods for the Solution of Ill-Posed Problems*, Kluwer, Dordrecht.

Tikhonov, A.N., Ryutin, A.A. and Agayan, G.M. (1983b) On a stable method for solving a linear programming problem with inexact data. *Dokl. Akad. Nauk*

SSSR, **272**, 1058–1063 (in Russian); English transl. in *Soviet Mathem. Dokl.*

Tikhonravov, A.V. (1982) The accuracy arising in principle when solving synthesis problems. *USSR Comput. Mathem. and Mathem. Phys.*, **22**, 143–156.

Trenogin, V.A. (1980) *Functional Analysis*, Nauka, Moscow (in Russian).

Vainberg, M.M. (1972) *The Variational Method and Method of Monotone Operators in the Theory of Nonlinear Equations*, Nauka, Moscow (in Russian).

Vaĭnikko, G.M. (1982) *Methods for Solving Linear Ill-Posed Problems in Hilbert Space*, Tartu University Press, Tartu (in Russian).

Vaĭnikko, G.M. (1988) On the regularization of ill-posed extremal problems, in *Numerical Methods and Optimization*, Valgus, Talinn, 56–65 (in Russian).

Vaĭnikko, G.M. and Veretennikov, A.Yu. (1986) *Iterative Procedures for Ill-Posed Problems*, Nauka, Moscow (in Russian).

Varah, J.M. (1973) On the numerical solution of ill-conditioned linear systems with applications to ill-posed problems. *SIAM J. Numer. Anal.*, **10**, 257–267.

Vasilyev, F.P. (1978) On the regularization of ill-posed extremal problems. *Dokl. Akad. Nauk SSSR*, **241**, 1001–1004 (in Russian); English transl. in *Soviet Mathem. Dokl.*

Vasilyev, F.P. (1980) *Numerical Methods for Solving Extremal Problems*, Nauka, Moscow (in Russian).

Vasilyev, F.P. (1981) *Methods for Solving Extremal Problems. Minimization Problems in Functional Spaces: Regularization, Approximation*, Nauka, Moscow (in Russian).

Vasilyev, F.P. (1988) Estimates for the rate of convergence of the Tikhonov regularization method for unstable minimization problems. *Dokl. Akad. Nauk SSSR*, **299**, 792–796 (in Russian); English transl. in *Soviet Mathem. Dokl.*

Vasilyev, F.P. and Kovač, M. (1984) On the regularization of ill-posed extremal problems with inexact initial data, Banach Mathem. Center Publications, Warsaw, **13**, 237–263.

Vasin, V.V. (1972) On the β-convergence of the projection method for solving nonlinear operator equations. *USSR Comput. Mathem. and Mathem. Phys.*, **12**, 266–274.

Vasin, V.V. (1975) The discrepancy method and finite-dimensional approximation of approximate solutions to operator equations, in *Methods for Solving Conditional-Posed Problems*, Ural's University Press, Sverdlovsk (in Russian) 39–52.

Vasin, V.V. (1979) Pointwise convergence and finite-dimensional approximation of regularizing algorithms. *USSR Comput. Mathem. and Mathem. Phys.*, **19**, 8–19.

Vasin, V.V. and Tanana, V.P. (1974) Necessary and sufficient conditions for convergence of projection methods for linear unstable problems. *Dokl. Akad. Nauk SSSR*, **215**, 1032–1034 (in Russian); English transl. in *Soviet Mathem. Dokl.*

Vinokurov, V.A. (1972a) An approximate discrepancy method for nonreflexive spaces. *USSR Comput. Mathem. and Mathem. Phys.*, **12**, 265–273.

Vinokurov, V.A. (1972b) Two notes on the choice of regularization parameter.

USSR Comput. Mathem. and Mathem. Phys., **12**, 249–253.

Vinokurov, V.A. (1979) On the error of the approximate solution of linear inverse problems. *Dokl. Akad. Nauk SSSR*, **11**, 792–793 (in Russian); English transl. in *Soviet Mathem. Dokl.*

Vinokurov, V.A. and Gaponenko, Yu.L. (1982) A posteriori estimates for solutions of ill-posed inverse problems. *Dokl. Akad. Nauk SSSR*, **263**, 277–280 (in Russian); English transl. in *Soviet Mathem. Dokl.*

Voevodin, V.V. (1969) On the regularization method. *USSR Comput. Mathem. and Mathem. Phys.*, **9**, 671–673.

Voevodin, V.V. (1977) *Computational Foundations of Linear Algebra*, Nauka, Moscow (in Russian).

Voevodin, V.V. and Kuznetsov, Yu.A. (1984) *Matrices and Calculations*, Nauka, Moscow (in Russian).

Wahba, G. (1977) Practical approximate solutions to linear operator equations when the data are noisy. *SIAM J. Numer. Anal.*, **14**, 651–667.

Wedin, P.Å. (1973) Perturbation theory for pseudoinverses. *BIT*, **13**, 217–232.

Wentsel', E.S., Kobylinskiĭ, V.G. and Levin, A.M. (1984) Applying the regularization method to the numerical solution of the problem of bending within elastic plates. *USSR Comput. Mathem. and Mathem. Phys.*, **24**, 202–205.

Wilkinson, J.H. and Reinsch, C. (1971) *Handbook for Automatic Computation. Linear Algebra*, Springer-Verlag, Berlin.

Yagola, A.G. (1979a) On the choice of regularization parameter by the generalized discrepancy method. *Dokl. Akad. Nauk SSSR*, **245**, 37–39 (in Russian); English transl. in *Soviet Mathem. Dokl.*

Yagola, A.G. (1979b) On the generalized discrepancy principle in reflexive spaces. *Dokl. Akad. Nauk SSSR*, **249**, 71–73 (in Russian); English transl. in *Soviet Mathem. Dokl.*

Yagola, A.G. (1980a) The solution of nonlinear ill-posed problems by the generalized discrepancy principle. *Dokl. Akad. Nauk SSSR*, **252**, 810–813 (in Russian); English transl. in *Soviet Mathem. Dokl.*

Yagola, A.G. (1980b) On the choice of regularization parameter while solving ill-posed problems in reflexive spaces. *USSR Comput. Mathem. and Mathem. Phys.*, **20**, 586–596.

Yagola, A.G. (1984) Ill-posed problems with approximately specified operators, in *Computational Mathematics*, Banach Mathem. Center Publications, Warsaw, **13**, 265–279 (in Russian).

Zagonov, V.P. (1987) Some variational methods for constructing approximations of nonsmooth solutions of an ill-posed problem. *USSR Comput. Mathem. and Mathem. Phys.*, **27**, 9–17.

Zaikin, P.N. and Mechenov, A.S. (1971) Some approaches to the numerical solution of integral equations of the first kind by the regularization method, in *Scientific Reports of Computer Center of Moscow State University*, Moscow University Press, Moscow, N 144–T3/463 (in Russian).

Zhukhovitskiĭ, S.I. and Avdeeva, L.I. (1964) *Linear and Convex Programming*, Nauka, Moscow (in Russian).

Index